THE KING'S OWN SCOTTISH BORDERERS

Also by Trevor Royle in the same series

The Royal Scots

The Black Watch

The Royal Highland Fusiliers

Queen's Own Highlanders

The Gordon Highlanders

THE KING'S OWN SCOTTISH BORDERERS

A Concise History

TREVOR ROYLE

MAINSTREAM PUBLISHING

EDINBURGH AND LONDON

First published in Great Britain in 2008 by
MAINSTREAM PUBLISHING COMPANY
(EDINBURGH) LTD
7 Albany Street
Edinburgh EH1 3UG

ISBN 9781845960919

A catalogue record for this book is available
from the British Library

Typeset in Bembo

Printed in Great Britain by
Clays Ltd, St Ives plc

Contents

Preface

In 2006 The King's Own Scottish Borderers underwent one of the many organisational transformations which have been visited on the British Army since its beginnings in the seventeenth century. As part of wide-ranging reforms to the structure of the infantry it amalgamated with The Royal Scots to form The Royal Scots Borderers, 1st Battalion The Royal Regiment of Scotland. This new 'large' regiment was one of several created throughout the army in 2006 and represented the most radical innovation since the completion of the Cardwell/Childers reforms in 1881 when single-battalion regiments were amalgamated and territorial names replaced the earlier system of numbering. Inevitably these latest changes created a great deal of sadness in the army community and more widely throughout Scotland, with regret being expressed for the loss of some cherished names and the conversion of regiments into a new formation. However, the history of the British Army shows that the story of its regiments has been one of constant development, with cutbacks, amalgamations and changes of name being part of a process of evolution stretching back over several centuries. In every case the development has not led to a diminution of the army's capabilities but has produced new regiments which are the equal of their predecessors.

Together with the other books in the series, this concise history has been written to mark this latest transformation in Scottish and British military history. This is not a new regimental history of the regiment and its predecessors, but I hope it will be a useful addition to the regiment's historiography. Invariably, as is the case with the other concise regimental histories in the series, it also reflects the history of the British Army and the empire in which it served. I owe a tremendous debt to previous regimental historians, whose books are listed in the bibliography. It goes without saying, I hope, that this history of the regiment could not have been attempted without a thorough reading of these volumes, especially the battalion histories which were written to commemorate the service of The King's Own Scottish Borderers in the two global wars of the twentieth century.

For help during the project and assistance with the selection of illustrations I would like to thank Lieutenant-Colonel Colin Hogg, the Home Secretary of The King's Own Scottish Borderers, and archivist Ian Martin. Grateful thanks are also due to Major-General John Cooper, who gave his blessing to the project during his period of office as Colonel of The King's Own Scottish Borderers.

Trevor Royle

ONE

Edinburgh's Own

In the spring of 1989, on a raw March morning, military helicopters swooped into Queen's Park in the heart of Edinburgh, and in the lee of Arthur's Seat started disgorging armed soldiers who assembled in the Abbey Close before marching up the city's Royal Mile, the main thoroughfare which runs from the Palace of Holyroodhouse to Edinburgh Castle. It was not the start of a military exercise and it most certainly was not a prelude to the annual military tattoo which takes place on the Castle Esplanade. There were no trews or kilts in sight, instead all the officers and men were wearing battle dress and carrying SA80 personal weapons. Rather, it was a dramatic and highly relevant way to commemorate the 300th anniversary of the foundation of the regiment which became known as The King's Own Scottish Borderers, and to reinforce the idea that it had come into being to defend the city of Edinburgh. As the Jocks marched through the streets of Edinburgh's Old Town the regiment was sending a message which echoed down the centuries: the Borderers were raised to defend the city and 300 years later they were still capable of carrying out that task, helicopters and all. For everyone associated with the regiment which has been closely associated with Scotland's traditional Border counties since the late nineteenth century, it was a proud moment as it was a deft and timely reminder that the Borderers have always enjoyed an intimate relationship with Scotland's capital city. In a very real historical sense the men of The King's Own Scottish Borderers, or KOSB, (but never Kosbie) were coming home.

Yet there is a paradox at the heart of the regiment's history. Its founding father was an ornament of the Scottish aristocracy; in its first years one of its titles was the Edinburgh Regiment; in time it came to be known for its connections to the Scottish Borders and for many years was counted as one of Scotland's most historic infantry regiments, third in antiquity to The Royal Scots and The Royal Scots Fusiliers. Everything about the modern regiment – its Leslie tartan, its cap badge including the Saltire and Edinburgh Castle – is representative of Scotland. For over 100 years it recruited its soldiers from the Border counties of the old East and West Marches, but for much of its life the regiment was not always considered a Scottish institution. Between 1782 and 1805 it was known as The Sussex Regiment and was numbered 25th in the army's Order of Precedence, with its regimental headquarters in the English town of Berwick-on-Tweed. Even when it became a royal regiment in 1805, having been granted the title of King's Own Borderers by George III, by the admission of one of the regiment's historians 'the ties with Scotland were fitful and feeble' and these were not consolidated until the radical reorganisation of the infantry in 1881 which ended single-battalion regiments by amalgamating them and introducing new territorial titles. It was a close-run thing: the 25th was to have been known as The York Regiment, King's Own Borderers, but following protests it received its Scottish title and has never looked back. It was right and proper that good sense prevailed because the 25th was founded in 1689 as a Scottish regiment to meet a distinct military need.

Besides, as Captain 'Sandy' Stair Gillon put it so admirably in his history of the regiment in the First World War, what counts is the spirit that pervades the regiment:

> An Englishman and a Scotsman may preserve their national characteristics and diversities to the grave. But put them together in the KOSB for a spell – and it doesn't take long – and both develop a second characteristic to be shared in common, namely, that of being Borderers. In that capacity the best of their fighting and enduring qualities are brought out and developed. The spirit is one very difficult to define. It is distinctly but unaggressively Scottish, and distinctly and possibly a little more unmistakably military. In short, it is just the child of the British Army, but a child with a very definite personality, which passes on from generation to generation and is absorbed by drafts and recruits with surprising rapidity.

Gillon knew what he was talking about. Not only did he serve in 1st KOSB but he was a close friend and climbing companion of John Buchan, a quintessential Scottish Borderer. (During the latter part of the war Gillon served as Buchan's associate in the wartime Department of Information which was established in 1916 to direct Britain's wartime propaganda initiatives.)

At the time that the regiment was raised at the end of the seventeenth century Scotland was going through one of its many periods of political and religious turmoil. Indeed, the ferment gripped the whole of the United Kingdom. Four years earlier, in February 1685, King Charles II had died and was followed to the throne by his brother James II. Initially the new reign got off to a promising start but soon James began a reckless policy of promoting Catholic allies, including the Earl of Tyrconnell as Lord Lieutenant in Ireland, Lord Jeffries as Lord Chancellor and the ambitious Earl of Sunderland as joint Secretary of State. Slowly, an inner circle of Catholic advisers ring-fenced the King and England became increasingly Catholic in complexion. A papal nuncio was received, new friaries and monasteries were opened in London, a Jesuit school appeared at the Savoy, the army welcomed Catholic chaplains and James began taking soundings amongst Justices of the Peace and deputy lieutenants of the counties about repealing the Test Act. A first Declaration of Indulgence ending laws against Catholics was passed in April 1687 (in Scotland the legislation appeared as Letters of Indulgence), and a second in May the following year. Similar steps were taken in Ireland to unstitch the Protestant ascendancy: Catholics were granted key political and legal appointments and the Irish parliament threatened to pass legislation to reverse the land settlement introduced by Oliver Cromwell. To the watching Irish Protestants the situation seemed ripe for a Catholic take-over.

The flashpoint was provided by the Church. In May James ordered the second Declaration of Indulgence (negating laws to punish Roman Catholics and Protestant dissenters) to be read in all churches; six bishops refused and were put on trial for seditious libel, together with the Archbishop of Canterbury. (They were acquitted amidst great public jubilation at the end of June.) With the King and the Church in direct confrontation, a group of conspirators made contact with William of Orange and invited him to intervene to save parliament and religion, 'if the circumstances stand so with your Highness that you believe that you can get here time enough, in a condition to give assistance this year'. (As the son of Charles II's sister Mary, William had a claim on the throne, as did his wife Mary, who was the daughter of James II.) This was the message James's son-in-law had been waiting for and he immediately began raising an invasion force. William's decision was prompted by one further event – the birth of a son and heir, James Edward Stuart, to his father-in-law and his second wife, Mary of Modena. It was inconceivable that the new prince would not be raised in the Catholic faith.

James attempted to make concessions, but it was too late and he knew it. Following the bishops' acquittal James had been surprised to see his soldiers cheering and was equally astonished when an officer told him not to concern himself. Things were falling apart and when William landed at Torbay on 5 November support began drifting away from the Crown towards the invading saviour. The magnates who had invited William began seizing the key cities, two senior army commanders the Duke of Grafton and Colonel John Churchill turned coat and with no army to defend him James was forced to escape to France. Even that proved difficult. The fishermen taking him uncovered his identity and brought him back, and it was not until the year's end that he was allowed to leave openly through the port of Rochester. The last Stuart king had departed and the dynasty was destined never to return to its three kingdoms.

The end of the Stuarts and the arrival of the Protestant house of Orange finally settled the relationship between the monarch, parliament and the people and in that sense the arrival of William and Mary put a gloss on the arrangements which followed the demise of the earlier Commonwealth and Protectorate. The new settlement was enshrined in the Declaration of Rights which was presented to William on 13 February 1689 together with the offer of the crown. This stated that parliaments must be held regularly, that elections must be free and that parliamentary debates must be open and not questioned in any place other than parliament, that only parliament could give consent to statutes, levy taxation and that 'raising and keeping a standing army within the kingdom in time of peace unless it be with consent of parliament is illegal'. The declaration also limited the succession to William's heirs and excluded Catholics. This time around, though, there was no bloodshed and the event was known to English history as the Glorious Revolution. Only in Scotland and Ireland was there violence, and this in turn led to the foundation of the regiment which would become The King's Own Scottish Borderers.

In Scotland not everyone was prepared to accept William as King and the opposition to the new order centred on a prominent Jacobite (as James's supporters came to be known), John Graham of Claverhouse, Viscount Dundee, who began gathering forces in his heartlands north of the Tay. At the same time the Duke of Gordon seized Edinburgh Castle in James's name. With rebellion and the threat of civil war in the air the Scottish Convention of Estates decided to take steps to raise a regiment numbering at least 800 men to protect the city. The order was encapsulated in an Act of Parliament dated 18 March 1689:

> The Committee doe offer it as of their opinion that for secureing the peace of the toune the Meeting of Estates may be pleased to grant warrand to the Earle of Levin [*sic*], with all expedition to levie ane Regiment of foot consisting of eight hundred men, and to beat Drummes to that effect. And that so soon as they are in readiness, he cause them Rendezvouze in the Abbey Close at the sight of, and that from the tyme they are Rendezvouzed there they shall be taken into pay.

Astonishingly, by beat of drum, the regiment was raised within four hours and numbered 1,000 men. It was known as the Edinburgh Regiment of Foot or, following the custom of the day, it was also named after its commanding officer, the Earl of Leven's Regiment. By any standards the Earl, David Melville, was one of the leading Scots of his day. Through his mother Catherine he was the grandson of Alexander, Lord Balgonie, the eldest son of Alexander Leslie, 1st Earl of Leven and one of Scotland's most distinguished soldiers. Leslie had served in the army of King Gustavus Adolphus during the Thirty Years War and on his return to Britain he had commanded the Scottish army, which had fought alongside the Parliamentary army in the defeat of King Charles I's Royalist forces at Marston Moor in 1644. Born in Fife in 1660, David Melville had succeeded to the Leslie earldom through his maternal grandfather in 1688 and was a committed Williamite supporter. He was also an experienced soldier in his own right, having served in the army of the Elector of Brandenburg.

According to the Royal Warrant which accompanied the foundation of Leven's regiment the establishment of the new formation consisted of the following officers and men: one colonel (Leven), one lieutenant-colonel (William Arnot), one major, nine captains, 12 lieutenants, 12 ensigns, one surgeon, one adjutant, one quartermaster, one chaplain, 24 sergeants, 36 corporals, 24 drummers and 780 private soldiers. Amongst the officers was a Huguenot, Captain Henry Verriere, who had accompanied Leslie in Europe. The speed with which the new regiment was formed sent an unmistakable message to James's supporters, reminding them that the Convention was prepared to back its policies with armed force. With Viscount Dundee now out of the central belt Edinburgh was safe for the time being. On 2 April Leven received permission to remove his regiment to new quarters in Fife before it joined the forces being raised by Major-General Hugh MacKay of Scourie to move against Claverhouse and his Highland host. Unlike Dundee, MacKay was a Highlander – he was the third son of Hugh MacKay of Scourie in Sutherlandshire and chief of the clan of that name – and unlike him he had no love of his fellow countrymen. Most of his life had been spent in foreign service in the Netherlands and according to the editors of his memoirs he had come to regard 'Scotsmen of those times in general as void of zeal for their religion and natural affection'. Three of MacKay's regiments had formed the Scottish–Dutch Brigade which had been serving under William of Orange in the Netherlands, but apart from the addition of Leven's new regiment he lacked manpower and he was determined to engage Dundee before the rebellion gathered strength. (In this and the following chapter the regiment will be referred to as the Edinburgh Regiment: following Leven's tenure as colonel army documents usually refer to it by the colonel's name or as the Edinburgh Regiment.)

The race was on to win support in the Highlands. MacKay offered huge bribes to try to convince the clans to join him but against that Dundee enjoyed the support of clan chiefs like Cameron of Lochiel and MacDonald of Keppoch, both of whom provided him with armed retainers. Having decided to base himself in Badenoch and Lochaber he slowly built up Jacobite strength and by the end of May had at least 1,700 men under his command. At the end of the month Dundee attacked Ruthven Castle and burned it down in an attempt to lure MacKay into battle but the commander of the Convention army refused to take the bait. Instead he kept his forces well clear of the Highland area and awaited developments. They were not long in coming. On 22 July Dundee moved out of Lochaber and advanced towards Blair Castle to confront MacKay's army, which was approaching nearby Dunkeld.

The long-awaited clash between the two armies came on the afternoon of 27 July in the Pass of Killiecrankie, through which the main north–south road runs. To MacKay's discomfort Dundee arrived first and occupied the high ground on the south-west slopes of Creag Eallaich, a move which gave the Convention army little chance of threatening the Highlanders' flanks. It also obliged MacKay to adopt a defensive posture by arranging his battalions in a long line three-deep with the Edinburgh Regiment on the right flank close to the policies of Urrard House. Behind them was the fast-flowing River Garry. Outnumbered by over two to one, Dundee knew that his best chance of success lay in the traditional Highland charge to cut through the opposition lines and sow confusion in the ranks. In the late afternoon the attack began and despite the intervention of MacKay's artillery pieces and the cavalry screen provided by Lord Belhaven's Horse, the Highlanders smashed into the left flank, causing it to break and scatter. While the right flank held firm – the Edinburgh Regiment did particularly well in this respect – the tactics used by Dundee effectively ended MacKay's control of his force. Realising that the battle was lost, he led the rump back to Stirling and safety.

It was an ignominious setback, but it was also a victory of sorts. During the attack Dundee was mortally wounded and his death effectively ended the rebellion. Shorn of his inspirational leadership the clansmen started drifting home and those who remained suffered a defeat at the hands of William Cleland's Cameronian regiment in the streets of Dunkeld later in the summer. Although Killiecrankie was counted as a defeat for the Convention army Leven's Regiment was one of the few formations to behave with credit and their service was recognised by a fulsome report which appeared in the *London Gazette* on 30 July:

> The Fight began about five in the Afternoon [it was in fact probably later] and was very sharp for some time; but some of our Regiments giving way, and Dundee's Men exceeding ours in number, they being about 6,000, part of our forces were put into disorder; whereupon Major-General MacKay thought fit to retire with the rest towards Stirling; where he arrived last night, with 1,500 Men, who retreated in a Body and in good Order: Of this number were the regiments of the Earl of Levens [*sic*] and Colonel

> Hastings, who as well as Officers as Soldiers, behaved themselves with extraordinary Bravery and Resolution; maintaining their ground to the last; and keeping the Field after the Rebels were drawn off to the hills.

For the Edinburgh Regiment the next two years were spent on garrison duty in the highland area, taking part in internal security duties to put down the remnants of Dundee's revolt. Their fate remained tied to the career of General MacKay, who was ordered next to take his army across to Ireland to join the Williamite forces led by the veteran Dutch General Godard van Reede de Ginkel. This campaign brought the Edinburgh Regiment into service in the second of the wars which followed the Glorious Revolution. Known as *Cogadh an Dá Rí* (War of the Two Kings), it followed James II's landing in Ireland at Kinsale in March 1689 and the subsequent armed resistance of the Protestants in Derry and Enniskillen. The armies of the two kings met at the Boyne on 12 July, which resulted in a serious defeat for the Jacobite forces but the fighting dragged on until 1691, due largely to the spirited leadership of Patrick Sarsfield. By the time that the Edinburgh Regiment entered the fray the French support had largely been withdrawn and Ginkel's army was in the ascendant. In Ireland the regiment fought at the siege of Ballymore and was part of the force which crossed the Shannon at Athlone prior to the decisive battle of the campaign – 'Aughrim's dread disaster' on the slopes of Kilmoddon Hill – where as Higgins's history records, the early death of the French commander ended all hopes of a Jacobite victory:

> After an obstinate engagement of two hours, wherein the Enemy lost their General, St Ruth, about 4,000 killed, 600 prisoners, and all their ammunition, artillery, tents, baggage

> and provisions, with twenty-nine stands of colours, twelve standards, and almost the whole of the arms of the infantry, which they threw away to expedite their flight; for having been totally routed, they ran off with the utmost precipitation, pursued by the British, who for four miles made a terrible slaughter amongst them.

Aughrim effectively ended Jacobite resistance in the *Cogadh an Dá Rí* and under the terms of the Treaty of Limerick 14,000 Irish soldiers were allowed to go into exile to become the mercenaries known as the 'wild geese' serving under the banner of the Irish Brigade in French service in the European wars which followed.

Once again the fortunes of the Edinburgh Regiment were to be tied to their commanding general. At the end of the fighting in Ireland Ginkel took his army to Flanders to serve in the Army of the League of Augsburg, a coalition formed by England, the United Provinces of the Netherlands, Spain and the German principalities to oppose King Louis XIV of France's expansionist policies in the Netherlands. At the time France was Europe's main power and there were fears that it would use its military superiority to gain hegemony over the continent and cow Spain into submission. Although the fighting in the Netherlands achieved nothing, it introduced the regiment to Britain's traditional campaigning grounds in Europe. In the coming years the Edinburgh Regiment would come to know Flanders and the Low Countries well, marching, counter-marching and fighting over ground with names that were to become familiar to generations of soldiers and, in time, part of the regiment's many battle honours.

War had been declared on 7 May 1689 and the first English army to serve in the theatre of war had been a force of 8,000 under the command of John Churchill who served under Prince George Frederick of Waldeck. It had secured an early victory over the French at Walcourt in August but the replacement of the French commander, de Villars, by the experienced Marshal François de Luxembourg changed the complexion of the fighting. The following summer he inflicted a crushing defeat on the allies at Fleurus and followed this up by taking Mons and routing Waldeck's army at Leuze. This was followed up in May 1692 by the capture of the fortress at Namur in a brilliant operation masterminded by Marshal de Vauban, the foremost military engineer of the day, who took only seven days to reduce the garrison. Attempts to raise the siege and save the Dutch–Spanish–German garrison were hampered by the persistent summer rains and the flooding of the tributaries of the River Meuse. It was at that stage of the war that MacKay arrived from Ireland with his battle-hardened regiments, including the Edinburgh Regiment, and it was not long before its men were in action under King William of Orange's overall (and, it has to be said, not very successful) command.

Following a series of attempts to engage the French army, all unsuccessful, the allies enjoyed a stroke of good fortune at the beginning of August. While on the march at night the allies unexpectedly came across Luxembourg's army at Steenekirk between the Scheldt and the Meuse, some 15 miles to the south-west of Brussels and to the south of the later and better known battlefield of Oudenarde. In the opening stages the French were surprised and their advance guard was quickly overwhelmed but Luxembourg quickly regrouped his forces and was able to draw them up in a new line of battle to meet the expected allied attack. To MacKay's eight regiments fell the duty of assaulting the French lines (most of their infantrymen were Swiss, then considered to be the best in Europe). It was not an easy task, as the attackers had to advance over difficult ground and when the actual fighting began it was marked by vicious close-quarter combat which lasted two hours before the French lines began to falter. Seeing that his men were being beaten back Luxembourg launched the French Guards, who were able to stem the breach opened by MacKay's men.

At that crucial stage in the battle, when the allies had gained an advantage which now had to be protected, William lost control. Instead of reinforcing MacKay's position he followed the advice given to him by the Dutch General Hendrik Solms, who had replaced Churchill. For reasons that are unclear and make no sense given the tactical situation, Solms decided to withhold the bulk of his army and left MacKay's regiments to face unequal odds. His only comment was chillingly unhelpful: 'Now we shall see what the bulldogs can do!' The comments of the other protagonists were equally contrary. As the Edinburgh Regiment and the other seven regiments began a fighting retreat from the French lines and men started falling in droves William burst into tears as he watched the slaughter and exclaimed, 'Oh, my poor English!' (Like many other things during this ill-starred campaign he was wrong on that count, as also taking part in the battle were The Royal Scots, the Scots Fusiliers Regiment of Foot and the Earl of Angus's Regiment, later The Cameronians.) As for MacKay, a pious man who believed that he was defrauding the Almighty if he was not doing his great work, he returned to the fighting with the fatalistic words, 'The will of the Lord be done.' He was the most senior of the British losses of the 620 officers and men who were killed or wounded at Steenekirk; at the time it was a bloody roll-call in a battle which the historian of the British Army, Sir John Fortescue, described as being 'admirably designed and abominably executed'.

Later in the month William, now with 70,000 men under his command, engaged Luxembourg once more at Landen (or Neerwinden) but enjoyed even less success. The French cavalry managed to break into the allied lines, which had been constructed with entrenchments and palisades along a stream called the Landen, a tributary of the River Geet between Brussels and Liège and, having broken the positions in the centre, forced William's men to retire from the battlefield. During the first part of the fighting the Edinburgh Regiment was part of the defensive line on the right flank together with the Scots Fusiliers and Angus's Regiment. Fortunately Luxembourg chose not to pursue the retreating army, otherwise William would have been forced to sue for peace. Around 20,000 of his army had become casualties, making the Battle of Landen the bloodiest battle fought in Europe for over 200 years. So great was the slaughter that the House of Lords recommended that no British general should ever again serve under the subordinate command of a Dutch soldier, whatever his rank.

However, the fact that William had not been conclusively defeated in the field encouraged him to continue the war despite the fact that it was a drain on the country's exchequer. It also made him more determined than ever to raise the siege of Namur, a formidable fortress which had been reinforced by de Vauban and was thought to be impregnable. Around 14,000 French troops under the command of Duke Louis de Boufflers formed the garrison and the strength of their position was confirmed when the opening rounds failed to make any impression on the defences. The defenders were helped by the fact that the outer works to the south were protected by the River Meuse and by the town, which lay to the east. Initially William ordered the operations to begin by attacking the heights at Bouge on 8 July and there were successive attempts throughout the month on 17, 23 and 24 July. The latter involved the Edinburgh Regiment and the difficulties involved in sustaining the effort can be seen from the account contained in Higgins's history:

> At four o'clock in the afternoon of the 27th July the English and Scotch, under the orders of Major-General Ramsay and Lord Cutts, issued from the trenches upon the right, and attacked the point of the advanced counterscarp which enclosed the great sluice of waterstop, near the gate of St Nicholas. In doing this they were terribly exposed to the fire of the counterguard and demi-bastion of St Roch; and the Enemy having exploded a mine under part of the glacis, whereby twenty officers and upwards of 500 men of Leven's were killed, some confusion ensued; but the troops having rallied, returned to the assault with redoubled vigour, when the Enemy gave way after a desperate resistance. The British now pursued them closely, and effected a lodgement on the foremost covered way, and on part of the counterscarp before the gate of St Nicholas; whereupon the Dutch, who seconded the assault, gallantly mounted a breach in the counterguard wherein they also made a lodgement, and by digging some traverses, both were enabled to maintain the positions they had gained.

It was not until 26 August that the French garrison was finally worn down by the incessant attacks and the need to defend the fortress walls and on 5 September the garrison finally surrendered the castle, marching out with only 5,538 of the original garrison of 15,000. During the operations the men of the Edinburgh Regiment were surprised to discover that the French had developed plug bayonets, which allowed them to fire their weapons while the bayonet was fixed. As a contemporary account explains, it was a rude awakening and led to the British Army developing a similar system, with MacKay of Scourie being given the credit:

> One of them [a French infantry regiment] advanced with fixed bayonets against Leven's, when Lieutenant-Colonel Maxwell, who commanded it, ordered his men to *screw bayonets* into their muzzles, thinking the enemy meant to decide the affair point to point; but to his great surprise, when they came within a proper distance, the French threw in a heavy fire, which for a moment staggered his men. Who nevertheless recovered themselves, charged, and drove the enemy out of the line.

The war was settled by the Treaty of Ryswick on 20 September 1697, and along with the rest of the British forces the Edinburgh Regiment returned home to be based in Edinburgh before moving to Fort Augustus and Fort William. (In addition to gaining a first battle honour at Namur, the regiment was immortalised in Laurence Sterne's novel *Tristram Shandy*, in which the comic character Uncle Toby claims to have served in Leven's while his servant Corporal Trim had been wounded at Landen.) For the most part the regiment resumed internal security duties but within a decade of returning to Scotland Leven's formed part of a force led by the Duke of Argyll which was formed to put down the latest Jacobite attempt to retrieve the British Crown for the House of Stuart. (During this period the regiment is also referred to as Shannon's Edinburgh Regiment, after its new colonel, Richard, Viscount Shannon.) The rising had been prompted by the death of Queen Anne in the previous year and the accession of George, Elector of Hanover, who claimed the throne through his maternal grandmother, Elizabeth, a daughter of James VI and I. With resentment growing in Scotland over the Act of Union of 1707 James Edward Stuart, the Old Pretender, decided to make a bid for the throne by appealing to Jacobite supporters in Scotland and the north of England but, like all the attempts made in the eighteenth century, it failed to find any widespread backing. In Scotland the military campaign of 1715 was led with great ineptitude by John Erskine, 6th Earl of Mar, who is known to history as 'Bobbing John' on account of his many prevarications and his failure to grasp the moment. He proved to be incapable of any decisive military action and according to the historian Andrew Lang, he was negligent in his planning: 'Mar seems to have regarded powder as a rare product of the soil in certain favoured regions, not as a commodity which could be made at Perth or Aberdeen by arts known to men.' Although Mar got off to a good start in the Highlands where support for the Stuarts was strong, raising the standard on the Braes of Mar and quickly capturing Perth, he failed to take any decisive action. Another uprising took place in northern England but it was equally hopeless, with the English Jacobites being quickly and easily defeated at Preston and as it turned out, only one battle was fought in Scotland. On the morning of 13 November 1715 the government and Jacobite armies confronted one another on the slopes of the Sheriffmuir Ridge north of Dunblane, Argyll with his men in two lines as follows (contemporary or founding titles are given together with post-1881 titles in brackets):

> First line, from left to right: two squadrons of King's Dragoons (later 3rd King's Own Hussars), two squadrons of Cunningham's Dragoons (later 7th Queen's Hussars), Hales Regiment of Foot (later The West Yorkshire Regiment), The Duke of Beaufort's Musketeers (later The Devonshire Regiment), Princess of Denmark's Regiment (later The King's Regiment) Royal Regiment of Welch Fuzileers (later The Royal Welch Fusiliers), Leven's Edinburgh Regiment, Wightman's Regiment of Foot (later The Royal Leicestershire Regiment), The Buffs (original title), two

squadrons of Princess Anne of Denmark's Dragoons (later 4th Queen's Own Hussars), two squadrons of Scots Greys (original title).

Second line, from left to right: one squadron of Cunningham's Dragoons, Royal North British Fusiliers (later The Royal Scots Fusiliers), Charlemont's Regiment of Foot (later 2nd Worcestershire Regiment), one squadron of Cunningham's Dragoons.

The battle began when Glengarry's Highlanders charged into the government army's left flank as it was still manoeuvring into position, a move which caused high casualties, but this was balanced by the charge of the Scots Greys on the right, who scattered the Perthshire, Angus and Fife horse, the only cavalry available to the Jacobites. By mid-afternoon Mar's numerically superior army had gained the upper hand but 'Bobbing John' failed to press home his advantage. Although he was left as master of the field his army was unable to march south while Argyll's regiments were able to retire towards Stirling. The stalemate at Sheriffmuir effectively ended the campaign, although it should be recorded that one of the officers in the Edinburgh Regiment, Captain the Hon. Arthur Elphinstone (later Lord Balmerino) resigned his commission and threw in his lot with the Jacobites. James Edward Stuart eventually landed in Scotland towards the end of the year, but by then support was ebbing away and there was great disappointment that he did not bring with him the expected French reinforcements. As a result Perth was abandoned and James went back into exile, first in France and then in Rome, which was to be his home until his death in January 1766.

For the men of the Edinburgh Regiment change was again in the air. Having taken part in the operations to put down the Jacobite threat they remained in Scotland until 1719, when they took part in another European conflict known as the War of the Quadruple Alliance. Following the death of Louis XIV in 1715 his grandson, Philip of Spain, entertained ambitions to gain the throne of France while his second wife, Elizabeth Farnese of Parma, pushed her own claims for territorial aggrandisement in Italy. This led to the creation of an alliance involving Britain, France and the Netherlands to oppose the Spanish claims and in 1718 it was joined by Austria. Most of the activity was confined to naval operations in the Mediterranean but in April 1719 a French army invaded the Basque provinces of Spain, followed by British amphibious landings along the Galician coast at Vigo and Pontevedra. In the latter operation, commanded by Lord Cobham, the Edinburgh Regiment was part of the force which captured the Spanish fort at Pontevedra, but this proved to be the highlight of the campaign and they were quickly withdrawn. During the attack on Pontevedra there was little resistance and the regimental records show that the only gains were 'some brass ordnance, small arms and other military stores'. On the regiment's return to Britain it proceeded to Ireland, first to Wicklow and then to Dublin. On 17 June Viscount Shannon was transferred to the Carabineers (later 6th Dragoon Guards) and was replaced as colonel by John Middleton. Later stations in Ireland were at Carrickfergus, Drogheda and Mullingar, and it was not until 1726 that the Edinburgh Regiment made its next move to the Rock of Gibraltar, which was to be the regiment's home for the next ten years.

During the deployment Gibraltar was besieged by Spanish forces, but not only was the garrison of 6,000 well defended, it was also well supplied with provisions brought in from north Africa and the records show that the British soldiers 'suffered but little loss, and treated the efforts of the Enemy with great contempt'. On 29 May 1732 the Earl of Rothes replaced Middleton as colonel and a return of officers seven years later (the earliest held by the regiment) shows that the Edinburgh Regiment consisted of the following:

Colonel: Earl of Rothes
Lieutenant-Colonel: James Kennedy
Major: – Biggar
Captains: James Dalrymple, David Cunningham, Henry Ballenden, Robert Armiger, John Maitland, Richard Worge, Lord Colville
Captain-Lieutenant: Frederick Bruce
Lieutenants: William Baird, Walter Brodie, George Scott, James Hairstreet, William Lucas, James Hamilton, David Watson, Archibald Douglas, David Home, Charles Steevens
Ensigns: James Livingston, George McKenzie, Thomas Goddard, James Sandilands, Robert Hay, Alexander Gordon, – Mackay, Thomas Goodrick, Patrick Lundie.

In 1736 all the private soldiers were transferred to Oglethorpe's Regiment and the officers and non-commissioned officers returned to Britain to recruit from scratch as tensions in Europe led to the expansion of the army by 20,000 men. It was fortunate that the regiment had a Scottish colonel in Lord Rothes, as it enabled the recruitment to take place in Scotland. The years of unbroken peace came to an end towards the end of the 1730s, when Spain emerged as a hostile threat to British trading interests, especially in the West Indies and Caribbean, and war seemed inevitable. To meet the need for more soldiers the regiment's ten companies were increased in size from the normal peacetime establishment of 34 private soldiers each to 70 private soldiers, and in 1740 they were further augmented by the addition of one lieutenant and 30 private soldiers. As a result of the emergency the regiment was deployed in the West Indies in 1740 but it was to be a short tour of duty. Soon they would be fighting once again in Europe.

TWO

Red Roses and Blue Bonnets

For most of the eighteenth century Britain's armed forces were on active service in the succession of wars against France. No longer was Spain the greatest threat (although it did not disappear altogether): following the accession of William and Mary of Orange in 1688 the French ambassador was ordered to leave London and warfare between the two countries was to be a periodic fact of life until the French army was finally broken and defeated at Waterloo in 1815. During that period the complexion of the British Army also changed. It was still very much the creation of the monarch and consisted mainly of 'subject troops' (soldiers enlisted in England, Ireland and Scotland) with 'subsidy troops' (mercenaries, mainly Danes, Hessians and Hanoverians) to reinforce them for the campaigns in Europe. There was also a Board of Ordnance which provided artillery, engineering and logistical support. At the beginning of the century, in the reign of Queen Anne, the numbers of subject troops ranged between 30,000 and 50,000 and the cost of maintaining them worked out at an average of £8 million a year (£1.65 billion in today's prices). Inevitably, there were huge reductions when campaigns came to an end, with wholesale disbandment of regiments, especially of those raised specifically for war service.

Recruiting was a perennial problem as the military life was widely unpopular and soldiering was often an employment of last resort. As the writer Daniel Defoe noted: 'In winter, poor starve, thieve or turn soldier.' Not that the army provided much in the way of recompense, with pay remaining at less than a shilling a day in the early years of the century. Because most of the recruits had criminal records punishment was severe – flogging and capital punishment were commonplace – and that hard regime did nothing to encourage shirkers to don uniform. Officers paid for their commissions and regiments were usually raised by wealthy landowners who had the necessary financial backing. All that the government supplied was a small lump sum and the basic equipment – cross-belts, swords, muskets and bayonets from the Royal Arsenals. As a result of the system infantry regiments were generally known by the surname or title of their colonel and, as we have seen, the Edinburgh Regiment was listed under several titles in the early eighteenth century as one colonel after another took over control of the regiment. On 25 April 1745 Rothes was moved to the colonelcy of the 6th Dragoons and was replaced by Lord Sempill, formerly colonel of the Highland Regiment of Foot (later The Black Watch). It was not until 1751 that the system was changed. Under a Royal Warrant of 1 July a Regular Establishment came into being with the cavalry, the foot guards and the line-infantry regiments being numbered in order of precedence – 14 regiments of cavalry, three regiments of foot guards and 45 regiments of foot. As a result the Edinburgh Regiment became the 25th Regiment of Foot. As to the colonels, they continued in name only but as the warrant made clear, numbers, not names, were to be the order of the day:

> No Colonel to put his Arms, Crest, Device or Livery on any part of the Appointments of the Regiment under his command.
>
> No part of the Cloathing [*sic*] or Ornaments of the Regiments to be Allowed after the following Regulations are put into Execution, but by Us [King George II], or our Captain General's Permission.

The changes meant that the five Scottish line regiments lost a lot of their national personality and with the exception of the one Highland regiment were often indistinguishable from their English and Irish opposite numbers. The other Scottish formations were 1st Royal Regiment of Foot (later The Royal Scots), 21st Royal North British Fusiliers (later The Royal Scots Fusiliers), 26th Regiment of Foot (later The Cameronians) and 42nd Regiment of Foot (later The Black Watch). All were to be involved in the fighting against France which dominated the rest of the century, and towards its conclusion they would also be joined by a number of Highland regiments which were later formed mainly for the campaigns in North America and India.

Although the 25th did not take part in Marlborough's campaigns and his great victories at Blenheim, Ramillies, Oudenarde and Malplaquet, in the opening years of the eighteenth century the regiment had already had its first experience of fighting against the French at Steenekirk and Namur. Its next opportunity, still as the Edinburgh Regiment, came in the conflict known as the War of the Austrian Succession (or King George's War) which, as its name suggests, was caused by a disputed claim to the Austrian throne. At the time of his death in 1740, the Emperor Charles VI of Austria left no male heir and was succeeded by his daughter, Maria Theresa. Although this possibility had been codified by the Pragmatic Sanction of 1713, a European convention which was supposed to guarantee the integrity of Maria Theresa's throne, the succession of a female ruler was opposed by Philip V of Spain, France and the Electors of Bavaria and Saxony. Britain was drawn into the conflict by Prussia's decision to invade the Austrian province of Silesia – Austria was a long-standing ally – and by France's threatening moves in Flanders. In the middle of 1743 a British force crossed over to Flanders under the command of King George II and was in action with its continental allies at Dettingen on 27 June 1743, the last time that a British monarch led his forces in battle.

Following a French advance into Flanders in the spring of 1745 George II's son, the Duke of Cumberland, took command of the coalition forces, which now numbered about 50,000 soldiers. Amongst them was the Edinburgh Regiment, which arrived shortly after Dettingen. The opening stages of the new campaign saw Marshal Saxe leading his French forces into Flanders to invest Tournai before taking up another defensive position at Fontenoy on the road to Mons. Saxe enjoyed numerical superiority and had ordered his men to construct an entrenchment system consisting of three redoubts on the hills to the east of the River Scheldt, but he was desperately ill with fever and was unwilling to hand over command to a deputy. Nevertheless, the opening stages of the battle seemed to go his way when the fighting began on 10 May with an ineffectual cavalry attack on the French defences. Cumberland's leadership style had been described by a fellow officer as 'outrageously and shockingly military', but there was little finesse in the initial moves, with the infantry attacking in three lines, the first two British and the second Hanoverian. To do this, on the following day, Cumberland adopted the tactics of the parade ground and led his men towards the French centre, where they dressed ranks and prepared to attack from within 50 paces of the opposition's defensive lines.

It was at that point that the army witnessed one of the great set-piece scenes of British military history. As the men waited for the order to begin the assault Lord Charles Hay, commanding the 1st Foot Guards (later The Grenadier Guards), stepped forward, took out a flask, raised it in a toast to the enemy and declared in stentorian terms: 'We are the English Guards and hope that you will stand till we come up to you and not swim the [River] Scheldt as you did the [River] Main at Dettingen.' (In that battle the Garde Français had given way and during the retreat many were drowned as they attempted to cross a bridge of boats.) To raucous cheers from both sides the battle began, with Cumberland's men firing a murderous fusillade which broke the French first line. At that Saxe ordered up his cavalry to stabilise the position, and for the next four hours it tried to break the huge allied defensive square, but to no avail. Amongst the most determined attacks were those made by the Irish Brigade, the 'wild geese' who had gone into French service following the crushing of the Jacobite cause in Ireland. Cumberland's only comment was pertinent and full of regret: 'God's curse on the laws that made those men our enemies.'

At the end of a day of hard pounding Cumberland was able to withdraw his army in reasonable order towards Brussels with neither side being able to claim any strategic advantage. British and allied casualties were reckoned to be around 7,500 killed and wounded, while the French losses were slightly lower at 7,200. The losses in the Edinburgh Regiment were one officer and 54 soldiers killed, eight officers and 76 soldiers wounded and 13 soldiers missing, presumed killed. Cumberland did not withdraw his forces until the end of the summer as Saxe took advantage of the situation to seize Tournai, Ghent, Bruges, Ostend and Brussels. Saxe's forces also laid siege to the town of Ath, which was garrisoned by the Edinburgh Regiment with the support of a number of Dutch detachments. The first attempt to take Ath failed but, suitably reinforced, the French returned at the end of September and the Dutch contingent decided to surrender in the face of superior odds. This left the Edinburgh Regiment, now reduced to 400 effectives, with no option but to follow suit; their only satisfaction was being allowed to march out of the town with all the honours of war.

After a short stay near Brussels Cumberland's army moved back to Britain, not just because the European campaigning season was coming to an end but because the regiments were needed to face a new threat posed by another Jacobite uprising. The claim was made by Prince Charles Edward Stuart on behalf of his father, who would have reigned as James III had the challenge been successful. Like so many other episodes associated with the Jacobite cause it had a romantic beginning and a tragic ending. On 25 July 1745 Charles, or 'Bonnie Prince Charlie' as he was soon to be known, landed in the western Highlands at Moidart accompanied by seven supporters and set about the near impossible task of encouraging the Highland clan leaders to join his father's cause. He received a mixed response. Some clan leaders supported him out of conviction and loyalty to the Stuart family: the first visible backing came from Cameron of Lochiel in Inverness-shire, who offered 700 loyal clansmen with the noble thought that, come what may, they would share the fate of their prince and leader. Others refused outright, including the Macdonalds of Sleat and the Macleods of Dunvegan, in whose lands the uprising had begun. Others hedged their bets by sending modest forces commanded by younger sons, but the sorry truth is that for all the historical romanticism engendered by the Jacobite Uprising of 1745–46, it was doomed from the start by the failure of the Highlands to give it widespread support and by the unwillingness of the French to provide any credible military backing. The Lowlands were another matter: by the middle of the eighteenth century they had lost any interest in the Stuart cause and were happier to embrace the benefits provided by the 1707 Act of Union which had joined together the parliaments of England and Scotland and laid the foundations for the modern British state.

Nevertheless, Bonnie Prince Charlie enjoyed a run of early successes which bolstered optimism amongst his supporters that his uprising might succeed, and with war still raging in Europe it caused chaos and crisis in London. It helped that the government garrison in Scotland was not only weak but badly led, and by the middle of September Edinburgh was in the prince's hands. The Jacobites' first military success came on 21 September at Prestonpans in East Lothian, where General Sir John Cope's government force was defeated and routed, having been unnerved by the speed and ferocity of the Highlanders' charge on their lines. After the battle a contemptuous song came into being, making fun of Cope's predicament – 'Hey, Johnnie Cope are ye waukin' yet?' – and it is one of the curiosities of military history that the tune is still popular in Scottish regiments as a reveille. Following that relatively easy success Charles's army marched into England with the intention of attacking London, but they never got that far. Discouraged by the onset of winter and the growing distance from their heartlands, Highlanders began deserting or making it clear to their officers that they had no stomach for further fighting. (At that time there was usually no campaigning during the winter months and Highlanders were especially prone to drift back home.) The lack of support from English Jacobites was also disheartening and, having reached Derby, on Friday, 5 December, Charles took the fateful decision to return to Scotland. Greatly depleted and with morale sinking Bonnie Prince Charlie's Jacobite army began the long march back to Scotland.

Even as they did so the government had already begun preparations to crush the rebels by employing superior force against them under the command of the Duke of Cumberland, freshly returned from his successes in Flanders. Amongst them was the Edinburgh Regiment (listed at the time as Sempill's), which crossed over to England in October to join an army which would eventually number 15 battalions of infantry, the irregular Argyll Militia, four units of 800 mounted soldiers and a small but effective force of artillery. Despite a setback at Falkirk at the beginning of the year Cumberland had the upper hand as far as resources were concerned. Not only was his force almost twice as large as the dwindling Jacobite army, but his troops were better trained and better equipped to withstand the shock of a Highland charge. When the two armies met on the open land of Drumossie Moor, between Inverness and Nairn, on 16 April on a raw early spring morning, there could only be one outcome. Despite the courage shown by most of the prince's men they were outnumbered almost two to one, they were badly led and most were tired, hungry and dispirited. In contrast Cumberland's men were confident of victory, cheering on their leader with shouts of 'Flanders, Flanders!' when he rode out to inspect them before the battle.

Cumberland had drawn up his infantry in two lines, with the Edinburgh Regiment on the left of the second line alongside Bligh's (later The Lancashire Fusiliers) and two formations of dragoons on the right (Kingston's and Cobham's Dragoons). The battle began around one o'clock, with the Jacobite artillery firing and the government army responding in kind and achieving far greater results in terms of casualties. Maddened by the incessant fire, the Jacobites made the first move when Clan Chattan led the charge followed by Fraser's Appin Stewarts and Lochiel's Camerons. Undaunted by the fearsome spectacle which had been their undoing at Prestonpans, Cumberland's front-line regiments stood firm and fired volley after volley into the unmissable targets. Hand-to-hand fighting followed as the surviving Highlanders careered into Barrell's (later The King's Own Royal Regiment) and Munro's (later The Royal Hampshire Regiment), which were directly in front of the left flank. Cumberland then ordered up the Edinburgh Regiment and Bligh's in support and, as he reported from Inverness on 18 April 1746, that settled the outcome of the battle which is known to history as Culloden:

> As their whole first line came down to attack at once, their right somewhat outflanked Barrell's Regiment, which was our left, and the greatest part of the little loss we sustained was there, but Bligh's and Sempill's, giving a fire upon those who had outflanked Barrell's, soon repulsed them, and Barrell's Regiment, and the left of Monroe's [*sic*], fairly beat them with their bayonets. There was scarce a soldier or officer of Barrell's, and that of Monroe's, who did not kill one or two men each, with their bayonets and pontoons.

During this frenzied stage of the battle Major-General John Huske, commanding the second line, rode up and ordered the men of the Edinburgh Regiment to remember the new bayonet drill. According to Colonel the Hon. Joseph Yorke, one of Cumberland's staff officers, 'He bid the men push home with their bayonets and was so well pleased that hundreds perished on their points.' The refusal of the left flank to buckle saved the day for Cumberland and on the right flank the belated charge of the Macdonalds of Clanranald, Keppoch and Glengarry was also checked by the steady fire of the Royals (later The Royal Scots) and Pulteney's (later The Somerset Light Infantry). Soon the fury of the battle slackened and with Cumberland's cavalry committed to the battle the Highlanders started casting aside their weapons and ran from the moor, leaving it to the smoke and the wind and the rain. Within an hour the subalterns in the government regiments were ordering their men to 'Rest on your Arms!', muskets were lowered and the men looked over the moor at the charnel house they had created. Even the veterans of Flanders were dumbfounded by the scene – a heaving mass of dead and dying piled up in the tufts of heather, the air thick with the moans and shrieks of the wounded. Amongst the rebels who did not flee was Lord Balmerino, formerly of the Edinburgh Regiment, who had switched sides after Sheriffmuir. Later he was executed for treason at the Tower of London.

During the fighting the Edinburgh Regiment's casualties were one killed and 13 wounded and as a result of their presence at Culloden they have the unique distinction of being the only British regiment to have taken part in the three Jacobite campaigns of 1689, 1715 and 1745–46. After the battle the regiment was spared the unpleasant duty of hunting down the survivors and implanting a policy which an officer writing in the *London Magazine* characterised as his men 'carrying fire and destruction as they passed, shooting the vagrant Highlanders they met in the mountains and driving off their cattle'. Instead the men marched back to Perth to join Wolfe's (later The King's Liverpool Regiment) and Pulteney's before proceeding to Burntisland for re-embarkation to Flanders. On arriving the three regiments were placed under the command of General Sir John Ligonier and took part in operations at Roucoux near Liège on 11 October 1746. More of a fighting retreat than a set-piece battle, the three infantry regiments formed the right flank with the Scots Greys and although vastly outnumbered the men gave a good account of themselves. In the aftermath of the fighting, in a letter to the Earl of Sandwich, Ligonier went out of his way to emphasise his belief that no credence should be given to French claims that the allies had been defeated:

> Three villages, occupied by eight Battalions, English, Hanoverians and Hessians, being attacked by fifty-four Battalions of French, and after repulsing them twice, were in their turn forced to give way; but the English Cavalry had all the advantage. I think that, properly speaking, this affair cannot be called a Battle, for I doubt if the third part of our Army was engaged. The cannonading was terrible for about two hours.

Roucoux was the last fighting of the campaigning season and the regiment went into winter quarters at Bois-le-Duc. On Christmas Day came the news that Lord Sempill had died and that he would be replaced as colonel by the Earl of Crawford. During the Battle of Roucoux he had commanded the second line of the cavalry, and while reconnoitring with another officer and his aide-de-camp had almost fallen into enemy hands. Only his quick thinking and his linguistic ability saved him. Before the French piquet could challenge him Crawford ordered them not to fire as 'we are friends' and asked them the name of their own regiment. On being told that they were the Regiment of Orleans, he replied: 'Very well, keep a good look-out; I am going a little farther.' At that Crawford spurred his horse and he and his little party made good their escape. Crawford remained as Colonel until December 1747, when he was replaced by the Earl of Panmure, who became the last named colonel of the regiment. He in turn was replaced by the Earl of Home in April 1752, a year after regimental numbers had been introduced.

Flanders was destined to be the regiment's home for the next three years as the war went into its next and ultimately inconclusive stage. At the beginning of July 1747 the Edinburgh Regiment took part in the Battle of Val (or Laffeldt), where it occupied the left of the allied line with Pulteney's and Munro's and a regiment of Hanoverians reinforced by artillery. During the fighting the Edinburgh Regiment captured two stands of French colours but lost one officer and 30 soldiers killed and four officers and 42 soldiers wounded or missing. The following year saw them engaged in the defence of Bergen op Zoom, one of the strongest Dutch fortresses, which was under siege by Saxe's forces for most of the summer. With its surrender the Edinburgh Regiment went into winter quarters at Breda and it was not removed from Flanders until 11 November 1748, after the war had been concluded by the Peace of Aix-la-Chapelle. Even then the regiment's troubles were not over. While sailing for the next destination, Ireland, the transports encountered heavy gales and were forced to take shelter at Harwich before continuing their voyage through the English Channel. Again contrary winds intervened and one of the ships was wrecked off the coast of Normandy. Although the men were saved they were obliged to march from Caen to Cherbourg, where a new ship awaited them. It was not until 27 April that the men eventually arrived at Kinsale, having left the Meuse almost six months earlier. In June the regiment marched to new stations at Cork and Cashel and remained in Ireland until March 1755, when it moved back to Scotland for the first time in 12 years. For the next year the 25th (as the Edinburgh Regiment had become) was stationed at Fort William and Fort Augustus, mainly on road-building duties, but the interlude also allowed the regiment to recruit and bring itself back up to strength.

The men of the 25th would soon be needed, because the country was on the verge of fighting a global conflict against the French which stretched from Europe to India in the east and the Americas in the west, as well as on the oceans in between. The Seven Years War, as it was known, was fought between Britain and Prussia on the one hand and Austria, Russia and France on the other, but in the mind of the British prime minister, William Pitt the Elder, it was a conflict aimed at destroying French colonial power or, in the words of Winston Churchill in his *History of the English Speaking Peoples* 'to humble the House of Bourbon, to make the Union Jack supreme in every ocean, to conquer, to command, and never count the cost, whether in blood or gold'. Initially it did not turn out that way. The conflict broke out in North America in 1755 with the French threatening the British colonies west of the Alleghenies, south from Louisiana and north from Canada, where they made rapid progress down the valley of the River Hudson. There were similar setbacks in India, where Calcutta was attacked and its inhabitants subjected to terrible imprisonment in confined quarters which came to be infamous as the Black Hole of Calcutta. These defeats were quickly reversed and by 1763 the Seven Years War had come to a successful conclusion through the Treaty of Paris, which gave Britain the upper hand over the French in North America, the West Indies and India, thereby establishing Britain as an imperial power.

Although the 25th did not participate in those colonial wars, it was involved in the campaigning in Europe. Again, the war began badly for the British, albeit amidst a number of successes won by the Prussians in Saxony. On 26 July 1757 Cumberland was on the receiving end of a heavy defeat at Hastenbeck by French forces under the command of Marshal Louis d'Estrées and as a result the allies were driven out of Hanover and forced to disband their armies through the terms of the humiliating Convention of Kloster Zeven. This was followed by two notable victories at Rossbach and Leuthen, where Frederick of Prussia's army scattered the opposition and secured the safety of both his own kingdom and Hanover. In their aftermath the Hanoverians, now under the command of the Duke of Brunswick, used the winter months to clear the French from the Rhine and thereby to protect Frederick's western flank. Brunswick's force was styled His Britannic Majesty's Army in Germany, and by the summer of 1758 it contained Hanoverians and Hessians, six regiments of British foot and 14 squadrons of British cavalry. Amongst them was the 25th, which crossed over to Europe that September together with the 12th (later The Suffolk Regiment), 20th(later The Lancashire Fusiliers), 23rd (later The Royal Welch Fusiliers), 37th (later The Royal Hampshire Regiment) and 51st Regiments of Foot (later The King's Own Yorkshire Light Infantry). Later they would be remembered as the Minden Regiments.

The battle of that name was fought on 1 August 1759 (it was also known as Thorhausen) and it brought undying fame to the British infantry regiments which took part in it. In overall command of the allied army was Prince Ferdinand of Brunswick, but the British infantry was formed in two brigades under the command of Major-General the Earl of Waldegrave (12th, 23rd and 37th) and Major-General Kingsley (20th, 25th and 51st). Lord George Sackville commanded the cavalry with the Marquis of Granby as his second-in-command and his forces consisted of 1st Dragoon Guards (later The Queen's Dragoon Guards), 3rd Dragoon Guards (later 3rd Carabineers), the Scots Greys and 10th Dragoons (later The Royal Hussars). The artillery pieces played a vital role in the battle and were under the command of Captains Phillips, MacBean, Drummond, Williams and Foy.

From the outset the odds were stacked against Brunswick, who had 80,000 men at his disposal compared to the two French armies of 66,000 and 31,000 under the overall command of Marshal the Marquis de Contades. The French also had the benefit of a superior tactical position, being drawn up to the right of the River Weser, with the town of Minden behind them, while the left flank was protected by marshy ground. In that position the French defensive posture could not be faulted – the cavalry was in the centre with the infantry on the flanks – but it was not particularly suited for an offensive operation and Contades was under pressure to inflict a decisive defeat on the allies. Meanwhile Brunswick drew up his army with the German forces on the right and the British and Hanoverian infantry on the left, the latter screened by Sackville's cavalry. For Contades any hope that the battle would go according to plan ended when the six British infantry regiments, backed by the Hanoverian Guards, suddenly and unexpectedly went into the attack. Later it was claimed that an order had been given that the infantry was to advance *on* the beat of drum but that Waldegrave misheard this as the infantry was to advance *to* the beat of drum.

Due to the alignment of both armies, which had been drawn up in two shallow arcs, the British infantry went into the attack not against its opponents on the French left but against the cavalry in the centre. Although they were joined by two additional Hanoverian infantry regiments, it still meant that nine infantry formations were taking on the might of the French cavalry, a situation that was unparalleled in warfare at the time. Fortunately the advance had not been noticed immediately, being shielded by a copse, but when the two brigades appeared with Waldegrave's in the van, the French immediately counter-attacked with 11 squadrons of heavy cavalry. When that failed to make any impression against the steady musket fire of the advancing regiments, a second attack was made with 22 squadrons, but this too was dispersed. A third charge was made and although it threatened to break the attacking force's flanks, it was beaten off by the sustained and disciplined firepower of the British and Hanoverian infantry. Later, an officer in the 12th Regiment of Foot wrote a short description of the action, which is contained in the 25th's records:

> Our British infantry, headed by Generals Waldegrave and Kingsley, fought with the greatest ardour and intrepidity, sustaining and repelling the repeated attacks of the Enemy with the most romantic bravery. The soldiers, so far from being daunted by their falling comrades, breathed nothing but revenge; for my part, though at the beginning of the Engagement I felt a kind of trepidation, yet I was so animated by the brave example of all around me, that when I received a slight wound by a musket-ball slanting on my left side, it served only to exasperate me the more, and had I then received orders, I could with the greatest pleasure have rushed into the thickest of the Enemy. Interest, honour, glory and emulation, conspired to render the Battle of Thorhausen [Minden] famous to posterity.

Now was the time to launch the British heavy cavalry to support Waldegrave's and Kingsley's men. With the French centre broken and their cavalry forces committed it would have been relatively easy to wrap up their line. On the allied left a French assault had been repulsed and the British artillery were causing havoc with heavy and accurate fire along the French positions. It was undeniable that the battle would have been won by a British cavalry charge but unaccountably Sackville ignored four orders given by Brunswick and prevented Granby from taking matters into his own hands. Later, Sackville was relieved of his command and was dismissed from the army only to re-emerge as Colonial Secretary at the time of the loss of the American colonies later in the century. That failure allowed Contades to remove his men from the field, albeit having lost 10,000 casualties, killed, wounded or missing. All the French commander had to say was: 'I never thought to see a single line of infantry break through three lines of cavalry ranked in order of battle, and tumble them to ruin.' The allied losses were 2,600, killed, wounded or missing, the majority being suffered by the right flank regiments. In the 25th the casualties were one sergeant and 18 men killed, seven officers, four sergeants and 115 soldiers wounded. A week later, the first report of the battle appeared in the *London Gazette:*

> The Army was at this time marching with the greatest diligence to attack the Enemy in front; but the Infantry could not get up in time, General Waldegrave, at the head of the British, pressed their march as much as possible; no Troops could show more eagerness to get up than they showed. Many of the men, from the heat of the weather, and overstraining themselves to get on, through morassey [marshy] and very difficult ground, suddenly dropped down on their march.

Minden occupies an honoured place in the history of the British Army. Not only did the attacking infantry regiments display steadiness and discipline while facing the successive heavy cavalry charges, but they triumphed in spite of being given the wrong orders. As a result, all six infantry regiments received Minden as a battle honour and each year Minden Day is still commemorated by the successor regiments with special parades when the officers and soldiers wear red roses in their caps or bonnets. The custom is a reminder that their predecessors plucked wild roses and placed them in their hats as a means of identification before they went into the attack. One other innovation came from the battle: according to the regimental records it was the first occasion when British infantrymen took aim before firing their weapons, which at that stage of development were capable, in skilled hands, of being fired at the rate of four shots a minute:

> The foregoing engagement is the first in which the British troops took aim, by placing the butt of the firelock against the shoulder and viewing the object along the barrel when firing at the enemy; in which mode they had been instructed during the preceding peace. On former occasions the firelock was brought up breast high and discharged towards the enemy a good deal at random; because it was considered a degradation to take aim, according to the present custom.

That same year saw the destruction of the French fleet at Quiberon Bay and command of the sea allowed the British to reinforce North America. As a result Montreal fell in September 1760 and at the beginning of the same year Sir Eyre Coote defeated the French at the Battle of Wandiwash to give the British the upper hand in India. At the same time Ferdinand of Brunswick kept up the pressure on the French and defeated them again at the Battle of Warburg on 31 July 1760. During the battle the Marquis of Granby retrieved the reputation of the British cavalry by leading a spirited charge during which he famously lost both his hat and his wig. The incident gave rise to the expression 'going at them bald-headed' and only added to Granby's military reputation. (Later, the British deployment to the Gulf in 1990–91 to oust Iraqi forces from Kuwait was known as Operation Granby.)

British forces remained in action in all the theatres of the war until peace came in 1763, with the 25th playing substantial roles at Warburg, Klosterkampen and Vellinghausen but the deployments came at a heavy cost. At the height of the fighting Britain had 203,000 soldiers, mainly German mercenaries, in its pay and the army had swollen to over 100 regiments. Clearly there had to be cutbacks and as part of the peace dividend it was agreed to reduce the army to 70 regiments and to fix the establishment at

17,000 soldiers for the home garrison (England and Scotland), 12,000 for Ireland, 10,000 for the colonies, 4,000 for Gibraltar and Minorca (the latter came into British possession as a result of the earlier War of the Spanish Succession, and this was confirmed by the Treaty of Paris) and 2,000 for the artillery. For the next four years the regiment was back in Scotland but from 1768 to 1780 it formed part of the garrison on Minorca. Being one of the regiments in the Mediterranean station the 25th was involved in the Siege of Gibraltar in 1782, a long-running operation which had come about as a result of France and Spain declaring war on Britain in 1779 during the American War of Independence (1775–83). The Great Siege of Gibraltar lasted four years but under the determined leadership of General George Augustus Elliott (later Lord Heathfield), the scion of a well-known Roxburghshire family, the garrison held out. During the fighting the 25th lost two men killed and nine wounded. Gibraltar was confirmed as a British possession by the Treaty of Versailles which ended the war in 1783. By chance the ship which transported the 25th to Gibraltar in September 1782 was HMS *Victory*, Nelson's future flagship at the great naval victory of Trafalgar 13 years later.

That same year, 1782, the 25th added 'The Sussex Regiment' to its title as a result of a War Office decision to associate English counties with various infantry regiments. At the time the regimental colonel was Major-General Lord George Lennox, whose brother was the 4th Duke of Richmond. As the family had strong links with the county, being substantial landowners, Richmond requested that his brother's regiment should henceforth be linked with Sussex through its military title.

THREE

The War against Napoleon

In 1789 France found itself facing one of the most profound changes to have affected any country in the modern age, one which would not be repeated until the triumph of the Bolshevik revolution in Russia 128 years later. Following the long succession of wars which had dominated most of the eighteenth century France had been impoverished, and successive kings had exacerbated matters with a multitude of repressive taxes and the preservation of privilege amongst the nobility and the clergy. When the people finally rose up against the conditions the reasons for their actions were many and varied, and it could even be said that matters had in fact been improved. The sovereign of the day, Louis XVI, was no worse than many of his predecessors, steps had been taken to attempt to reform the administrative muddle that was the French government and some of the worst abuses were being addressed, but in the summer of 1789 that counted for nothing when a National Assembly was proclaimed and revolt came to the streets of Paris. At first it was thought that the revolution had fulfilled its aims – it was widely welcomed in Britain and other parts of Europe – but within four years, again for reasons which are multifarious, the revolution had been hijacked by extremists. Leadership became the preserve of demagogues, the royal family was executed on the newly-invented guillotine and Revolutionary France was at war with the rest of Europe. Ahead lay 22 years of warfare which would see the rise of a new emperor, Napoleon Bonaparte, and his eventual defeat at Waterloo.

Against that background, Britain's armed forces had to prepare for war with the reduced equipment and personnel that had been bequeathed to them as a result of the cutbacks following the conclusion of the Seven Years War and the fighting in North America which had seen the loss of the American colonies in 1783. Some idea of the depth of the cutbacks can be seen in the records of the 25th. Following its involvement in the siege operations at Gibraltar the regiment was reduced to eight companies of 51 rank and file each and there was a further reduction to 40 rank and file in 1787. At the same time the navy had a smaller number of ships, the army had been denuded of men with a bare minimum of 50,000 effectives, but as Pitt told the House of Commons, Britain had no option but to fight the French both to defeat a dangerous adversary and to defend the national interest in the country's greatest hour of need:

> We are at war with those who would destroy the whole fabric of our Constitution. When I look at these things they afford me encouragement and consolation, and support me in discharging the painful task to which I am now called by my duty. The retrospect to that flourishing state in which we were placed previous to this war ought to teach us to know the value of the present order of things and to resist the malignant and envious attempts of those who would deprive us of that happiness which they

> despair themselves to attain. We ought to remember that very prosperous situation at the present crisis supplies us with the exertions and furnishes us with the means that our exigencies demand.

Those were brave words but, not for the first or the last time in the country's history, Britain lacked the military means of enforcing its will on the continent. If it had possessed an expeditionary force, or, at the very least, a large enough army capable of immediate deployment in France, a pre-emptive strike on Paris could have discommoded the opposition but such a force was not available to Pitt. Instead he had to make do with what he had, and that meant deploying some 5,000 men under the Duke of York to Flanders, where they were joined by 13,000 of the King's Hanoverians and some 8,000 Hessian mercenaries. Most of the men were untrained, ill-equipped and unsure of what they were meant to be doing and their efforts were only redeemed by their courage and forbearance, and by the willingness of their officers to learn lessons as the campaign progressed. Even so, the deployment was not a credit to British arms. An attempt to take Dunkirk ended in disaster and everywhere the forces of Revolutionary France were in the ascendant.

Fortunately for the 25th, at that stage of the conflict its men were not called on to serve in the Low Countries. Instead, while the main bulk of the regiment prepared for a deployment to the West Indies, detachments of the 25th became Marines or 'sea soldiers', a singular honour which the regiment shares with ten other regiments – the 2nd Foot (later The Queen's Regiment), the 10th Foot (later The Lincolnshire Regiment), the 11th Foot (later The Devonshire Regiment), the 29th Foot (later 1st Worcestershire Regiment), the 30th Foot (later 1st West Lancashire Regiment), the 49th Foot (later 1st Berkshire Regiment), the 69th Foot (later 2nd Welsh Regiment), the 86th Foot (later 2nd Royal Ulster Rifles), the 90th Foot (later 2nd Scottish Rifles) and the 95th Rifle regiment (later The Rifle Brigade). As a result of those deployments the 25th had its establishment increased by ten men per company on returning to England from Gibraltar in April 1792 and within a year it numbered 600 rank and file, a substantial increase. As the regiment was based at Plymouth it was natural that it should have been selected to augment the navy's Marines for service on board the fleet of capital ships which were preparing for service under the command of Admiral Lords Howe and Hood. Captain Higgins's regimental records give a good picture of the process which took the 25th into the service of the Royal Navy:

> Captain H. A. Wright, Ensign Hamlet Wade and B. H. Hemings, with three sergeants, two drummers and seventy-four rank and file, embarked on board His Majesty's ship the Boyne, of ninety-eight guns, on the 25th of February. And on the same day Captain George Smith, Lieutenants S. V. Hinde and Haviland Smith, with four sergeants, two drummers and ninety-eight rank and file, embarked on H.M.'s ship the St George, of ninety-eight guns; and in a few days after another detachment, consisting of Captain John Stewart, Lieutenant Ascott Bickford, and Ensign George Vigoureux with three sergeants, two drummers, and seventy-four rank and file, embarked on H.M.'s ship Egmont, of seventy-four guns.

Soldiers had served at sea as Marines since the evolution of the country's armed forces in the previous century but, like many other arms, they had been cut back to parlous levels. Of the 16,000 men available to the Royal Navy in 1793, only 4,500 were Marines and these were insufficient for defence of the main ports and for manning warships at sea. Although Marines had no part to play in the operation of ships they had a vital role to play in battle – from providing firepower from the poop, quarterdeck and forecastle to repelling boarders. They also provided the manpower for boarding parties and were used in landing operations, where they fulfilled the same function as infantry. Following the successful use of Marines during the amphibious operations at Belle Isle during the Seven Years War, when two battalions joined nine infantry battalions to capture the French possession, the number of Marines was increased to 19,000. Marines also took part in many of the operations during the American War of Independence, fighting both ashore and on the warships operated by the Royal Navy.

As a result of the need to provide similar protection in the war against Revolutionary France the sea-soldiers were soon involved in actions which would have a bearing on the course of the war. In fact, the first of these would also influence the future course of world history. Although the men did not know it at the time, their first encounter with the French involved them with the man whose personality and leadership qualities would change the direction of the war – Napoleon Bonaparte, a young Corsican artillery officer, who would later become the Emperor of France. The engagement took place at Toulon in the summer of 1793, after the town and its naval port had declared its loyalty to King Louis XVII. Marseilles and Lyons followed suit but the revolts there were quickly put down, leaving Toulon in Royalist hands but desperate for reinforcement from the powers opposed to France. Britain's response was to send in Admiral Hood's Mediterranean fleet, but his 22 warships could not hold the heavily fortified port unaided and Hood made an immediate appeal for the despatch of 50,000 troops to mount the shore defences. Desperate attempts were made to find the necessary forces but with the British Army's manpower resources in a parlous state and its available troops deployed in the

West Indies and in Flanders under the Duke of York, it proved to be an impossible task. The Gibraltar garrison was shipped in and together with a mixed force of French Royalists, Neapolitans, Piedmontese and Spaniards they formed the allied army for the defence of Toulon.

For a time it seemed as if the defenders might have been able to secure the port. The French had to deal with the threat of Austrian invasion and the presence of the British expeditionary force in Flanders but the Committee of Public Safety, France's ruling body, dealt with each threat in turn before turning its attention to Toulon. Force of numbers decided the issue – the defending garrison numbered only 12,000, the French besiegers were three times that number – but the attack was also helped by Napoleon Bonaparte's astute handling of the artillery at Fort l'Aiguillette, which guarded the entrance to the harbour. During the operations detachments of the 25th were in action at Olioulles on 30 August, the heights of De Grasse on 30 September and the heights of Pharon the following day. The British force bravely attempted to hold the perimeter but on 17 December the defences broke and the defenders had to withdraw to the safety of Hood's fleet. Desperate attempts were made to scuttle the French fleet and blow up the port installations, but while these were only partially successful Hood's warships were at least able to withdraw safely, taking with them their detachments of sea-soldiers. From Toulon the fleet moved to Elba prior to an attack on Corsica by a small force under the command of Lieutenant-Colonel John Moore, an enterprising officer from Glasgow who was destined to make his name, and lose his life, in the later fighting against Napoleon in Egypt, Spain and Portugal. In the first action against French defences at Martello Bay on 17 February 1794 Moore commanded the attacking columns and his successful action was a prelude to the capture of the towns of St Fiorenza and Bastia.

Following the capture of Corsica the 25th returned to sea and detachments were in action again in March 1795 when Admiral Hotham's fleet made contact with the French in the Gulf of Genoa. They played a useful role in helping to secure the French ships *Le Ça-ira* and *Le Censeur*, which were seized for their prize-money. Following his success Hotham offered support to the Austrian forces operating along the coast between Genoa and Savona but he was forced to retire when Bonaparte's forces gained the upper hand. Those on board HMS *St George* experienced an unusual happening in July 1797 when they assisted the ship's officers in quelling an intended mutiny. It was not until October that the Mediterranean detachments returned to Britain, having also served in operations against the Spanish fleet off Cape St Vincent on 14 February. As Higgins's records make clear, the men of the 25th returned home with their pockets full:

> The St George returned to England, and on the 9th of October the detachment landed at Portsmouth, and marched to Plymouth Dock, where it joined The Regiment on the 4th of November. The men had received so much prize-money, that they paid for ringing the bells in every town through which they passed. Nevertheless no irregularity or want of discipline occurred during the march. By the capture of the St Iago Spanish galleon, the Captains of the 25th Regiment present received about one thousand five hundred pounds prize-money, and the Subalterns about nine hundred each.

Once back on dry land the men could exchange accounts of their experiences, one of the most eventful being the regiment's participation in Lord Howe's great naval victory over the French 400 miles off Brest, known as the 'Glorious First of June'. They were amongst the 18 army detachments which served on board British warships during this comprehensive victory over a larger French force. In fact the battle did not achieve its objectives – Howe was attempting to intercept a grain convoy from North America – but it provided a welcome victory, as the French fleet under Admiral Villaret de Joyeuse was forced to break off and seven enemy ships were captured or destroyed. During the battle the 25th's casualties were five soldiers killed and 14 wounded on HMS *Marlborough* and one killed and two wounded on HMS *Gibraltar*. For the British it was a welcome fillip. Although 130 French merchant ships managed to avoid Howe and sailed into Brest the British seamanship had been superior. Interestingly, although there has always been a traditional rivalry between the different branches of armed services, the regimental records show that the 25th had utter faith in their nation's naval superiority and was prepared to give credit where credit was due: 'The French seamen were of the ordinary sort; but their officers were, to all appearance, very inefficient and disreputable, being generally of all descriptions and ages, and taken from the second class of masters of merchant ships; their midshipmen also were anything but young gentlemen.'

On returning to Britain at the end of the naval deployment the remaining detachments were used to strengthen the rest of the regiment which had deployed to the West Indies in February 1795. To meet the need for recruits a 2nd battalion was also raised in Plymouth that same year in 1795 but it was quickly incorporated in the 1st battalion as the regiment found to its cost that its men had not been sent to a healthy station. In the 20 years between then and eventual victory in 1815, some 80,000 British soldiers served in the West Indies and half that number succumbed to illness, mainly yellow fever, leaving Major-General John Strawson, a recent historian of the period, to complain that 'no remote sugar island was worth such a price when the key to success against France lay not in the West Indies but on the continent of Europe'. So it proved for the 25th. Its first base was Grenada, where the main problem turned out to be local 'brigands' or insurgents supporting the French. In the first action against their positions on Pilot Hill the regiment sustained 67 casualties but these losses were outweighed by the scourge of disease. By the end of October the 25th had lost 259 officers and men and would lose many more before the deployment came to an end.

A worse fate awaited one company while it was being transported to the West Indies on board the transport *Belfast*. It was overtaken by a French corvette, the *Decius*, and the officers and men were taken prisoner and clapped in irons. Nothing daunted, they attempted to break loose but were foiled when their plan was revealed to the enemy. Transferred to other vessels when the corvette reached St Martins the men eventually had their revenge at Guadeloupe, where they succeeded in overpowering their captors and forced the ships to sail to Grenada. Once in safety they were able to rejoin the regiment. However, the officers had been isolated and remained in French captivity for ten months when they faced conditions described in the records as 'every degree of privation, filth and wretchedness, having no allowance or advantages above the common felons; they were dragooned, like them, from the deck and fresh air at stated hours, by Negro *citoyen* soldiers at the point of bayonet'. An exchange of prisoners saved them from certain death either from the privations caused by their confinement or from the ever-present threat of disease.

Some idea of the problems facing regiments in the West Indies can be found in the fact that it took 11 years to repair the condemned barracks at Orange Grove in Trinidad and 20 years to build new barracks at Fort Charlotte in the Bahamas. In an attempt to address this problem – for years the Caribbean had long been a death-trap for the regiments of the British Army – the War Office introduced a regular rotation of regiments. Shorter tours of duty were ordered for the most insalubrious spots and a new pattern of service was introduced whereby regiments were posted to the Mediterranean for acclimatisation before being posted to the heat and humidity of the West Indies or North America. Later this would be extended to the eastern hemisphere, where regiments spent time in Australia or South Africa before proceeding to India or Ceylon (now Sri Lanka). Between 1839 and 1853 the British Army suffered 58,139 casualties to disease or illness, and contemporary War Office papers reveal that the annual death rates per 1,000 men were 33 for non-commissioned officers and men and 16.7 for officers (in Jamaica it was 69 per 1,000 for officers and men). At the same time steps were taken to reduce the burden on the British Army by raising colonial corps and cutting back the number of garrisons in colonies which, it was reasoned, could just as easily be protected by the new generation of steam-powered warships of the Royal Navy.

In the summer of 1796 the 'miserable remains' (the description in the regiment's records) of the 25th returned to Plymouth, where it brought itself up to strength again and the regimental records from 1808 to 1816 show that the make-up of the soldiers during this period was as follows – 344 Scots, 613 English, 506 Irish and 37 Foreigners, 14 of whom were Caribbeans who had stayed on with the regiment as officers' servants following its recent deployment in the area. When the 2nd battalion was re-raised for service in 1804, its record book shows that it consisted of 576 Scots, 639 English, 288 Irish and eight Foreigners, probably Hessians or Hanoverians. In the case of both battalions some of those listed as English might have been Welsh but the figures give a good indication of the problem of recruiting soldiers in Scotland. Most of the Scots came from the central belt and the Border counties, while the numbers of English and Irish reflected the army's reliance on both those countries for a steady supply of recruits. The demography of the army during this period is also revealing: in 1830 there were 42,897 Irish non-commissioned officers and men while in the same year the figure for Scotland was 13,800. England and Wales provided a narrow majority with 44,329.

The return to home quarters was not uneventful. On arriving in Plymouth the regiment witnessed the explosion of the frigate HMS *Amphion*, which blew up in the dockyard with considerable loss of life. The following year there was a mutiny in the home fleet which affected crews at Plymouth, Portsmouth, the Nore, Sheerness and the North Sea. Attempts were made to involve the resident battalions but apart from unrest amongst a Royal Artillery detachment at Woolwich the army remained loyal and there was no disaffection in the local garrisons. To make the position of the 25th perfectly clear the regiment's non-commissioned officers issued a declaration of loyalty headed with the regimental motto *Nemo me impune lacessit* and warning troublemakers that they would do everything in their power to bring them to justice:

> The subscribing non-commissioned officers of His Majesty's 25th Regiment of Foot find with great regret that attempts have been made, by base and infamous persons, to alienate some of the soldiers in this garrison from their duty to their King and Country by circulating inflammatory papers and handbills, containing the grossest falsehood and misrepresentation, thereby insulting the character of a British soldier. In order to bring such incendiaries to the punishment they so richly deserve, we hereby offer a reward of ten guineas (to be paid on conviction) to the person or persons who will inform upon, secure, or deliver over to any of the subscribers, the author, printer or distributor of papers or hand-bills criminal, to the military establishment and laws of the country; or for any information against any

> person found guilty of bribing with money, or of holding out any false allurements to any soldier in this district, tending to injure the good order and discipline of the Army; which reward of ten guineas is raised and subscribed by us for this purpose, and will immediately be paid on conviction of any such offenders.

The next station for the regiment was the island of Jersey, where the garrison also included the 49th Foot (later 1st Berkshire Regiment), the 69th Foot (later 2nd Welsh Regiment) and the 88th Foot (later 1st Connaught Rangers). In August 1799 the 25th returned to its old campaigning ground in the Low Countries when it was ordered to join an expeditionary force of 10,000 soldiers under the command of General Sir Ralph Abercromby to encourage the House of Orange to throw in its lot with the allies against Revolutionary France. The initial stages of the campaign achieved early success, with the speedy deployment of the allied army at Den Helder and the equally rapid capture of the Dutch fleet off Texel Island, but thereafter matters did not run so smoothly. There was a lack of cooperation between the British and Russian field commanders and this was not helped by the death in action of the Russian commander, General D'Hermann, the fact that the Dutch were indifferent at best and hostile at worst to the arrival of their supposed liberators and there was also a worrying shortage of artillery pieces. The one major battle involved a frontal attack on the French positions at Egmont-op-Zee on 2 October when the 25th was in the vanguard of the British advance during the fighting on the sand dunes of the Zuyder Zee. The battle itself was inconclusive, with the French simply retiring towards the line Wyk–Kastrikum–Akersloot but as the regimental records show the 25th enjoyed a fair degree of success in fighting which was confused and lacking in co-ordination:

> Major-General Moore's Brigade which formed the right of the Army in column upon the beach, had The 25th Regiment as its advance guard, which was connected with the next column by means of a rifle company of The 60th Regiment extended across the sand-hills. General Moore moved forward without interruption until ten o'clock, when he approached the French drawn up under cover of a ridge of sand hills . . . Having received the enemy's fire, the brigade charged, by order of the General, and instantly drove the enemy from the first ridge; but in effecting this it suffered severely and The 25th Regiment had two officers killed and one mortally wounded. The contest between General Moore's Brigade and the Enemy continued very warm from ten o'clock in the morning till four in the afternoon, the nature of the ground enabling the French to dispute every inch.

The losses to the 25th at Egmont-op-Zee were two officers and 34 soldiers killed, eight officers and 63 soldiers wounded and 13 soldiers missing. During the battle the regiment was under the operational command of John Moore, now a major-general commanding 4 Brigade. Their efforts and those of the supporting British regiments were hindered by the use of *tirailleurs*, skirmishing French sharpshooters whose accuracy of fire and speed of movement caused high casualties amongst the advancing British redcoats. Similar troops known as *Jäger* had served in the Austrian and Prussian armies earlier in the century and had fought under British command in America, and the usefulness of these light troops was not lost on Moore, who was himself wounded during the fighting. Three years later he was instrumental in developing an Experimental Rifle Corps which was given instruction in light infantry tactics, field craft and shooting with the new Baker rifle which had greater range and accuracy than the contemporary smooth-bore musket.

In common with the other regiments who came under Moore's command, the 25th was fortunate to be associated with one of the great innovators in the British Army. Still a young man – he was below the age of 40 – Moore was very much a soldier's soldier who believed in the value of training and always put the needs of his men first. In common with other great leaders he argued that all ranks should share the privations and dangers of service in the field and he was insistent that soldiers in authority should not order their men to do anything unless they were also prepared to carry out the same duty. Above all, he was committed to the regimental system, seeing unit cohesion as the best means of maintaining morale and instilling discipline. 'It is evident that not only the soldiers but that each individual soldier knows what he has to do,' he remarked after inspecting his old regiment, the 52nd Foot (later 2nd Oxfordshire Light Infantry). 'Discipline is carried on without severity, the officers are attached to the men and the men to the officers.'

The 25th's next contribution to the war was to be equally bloody, but unlike Egmont-op-Zee it was crowned by success. This was Sir Ralph Abercromby's expedition to engage Napoleon's army of the east in Egypt in 1801, and once again the regiment was under Moore's operational command. Following the earlier meaningless operations in the Mediterranean, which had involved the 25th as sea-soldiers, the intention was to oust French forces from Egypt and to relieve the threat which they posed to Britain's holdings in India. However, the operation was not without hazard. Not only were the French already in position at Alexandria but they had more artillery and possessed cavalry. They were clearly in a good position to oppose the amphibious landings, but thanks to strict training in advance of the operation the British force came safely ashore at Aboukir Bay on 8 March and quickly formed a beach-head, forcing the French to withdraw. The main battle took place on 21 March at Canopus between Aboukir and Alexandria and it was a ferocious business, with the French losing at least 4,000 casualties and the British half that number, one of whom was Abercromby. Under Moore's operational direction the defending British forces showed great coolness under fire. Although the 25th was not involved in the first phase of the operations it provided reinforcement during the summer and landed in Egypt in August in time to take part in the operations to capture Alexandria, which duly fell on 3 September. To recognise their courage all the regiments involved in the expedition (including the 25th) were granted the right to bear on their colours the figure of the Sphinx, superscripted with the word 'Egypt'. As a result of the victory the French were evicted from Egypt and there followed a temporary and ultimately unsatisfactory truce through the Peace of Amiens, which was negotiated in the winter of 1802–03.

This arrangement was by no means perfect but at least it gave Britain a much-needed breathing space. It also provided Napoleon with the opportunity to rethink his strategy by courting Spain in order to gain possession of Louisiana in North America and Elba and Parma in the Mediterranean. He also laid claim to Malta, which had been captured by Britain in 1800 and was a vital strategic base for controlling the Mediterranean. However, having cowed most of Europe – Prussia, Russia and Austria were all defeated in 1803 – Napoleon's ultimate aim was to invade Britain with an army 200,000 strong which included his most experienced field commanders. It was a moment of supreme danger but in October 1805 the enterprise was foiled by Admiral Lord Nelson's famous victory at Trafalgar, where the French and Spanish fleets were destroyed. Even so, Britain's defences were still in a parlous state. Acting on the belief that the Treaty of Amiens had settled the nation's security the government under Prime Minister Henry Addington had introduced savage cuts in the armed forces in order to pay for the war. The army had been halved, with its strength set at 95,000 together with a garrison of 15,000 in Ireland and a part-time militia of 50,000.

There had to be a complete rethink of military policy when Britain declared war on Napoleonic France in May 1803. As had been forecast, the truce was shortlived and once more the army had to expand to meet the threat of invasion and the longer-term aim of defeating Napoleon's Grande Armée, which numbered 200,000 soldiers in seven army corps, commanded by soldiers of the calibre of Bernadotte, Marmont, Davout, Soult, Lannes, Ney and Augereau. To meet the need for additional soldiers regiments were given permission to raise further battalions and, as we have seen, the 2/25th came into being in 1804 and was used mainly for supplying drafts to the 1st battalion. There was a further change for the regiment a year later when it was made a royal regiment, a singular honour which allowed the men to wear blue facings on their uniforms. The regimental colour was also changed from buff to blue to reflect the change in status; the Sussex title was removed and the regiment became the 25th (King's Own Borderers) Regiment of Foot. At the time the 25th was based in Ireland, having been based for a short time in Gibraltar at the conclusion of the fighting in Egypt.

Unlike the other Scottish infantry regiments of the day, the 25th did not play any part in the Peninsula and Waterloo campaigns which saw the final defeat of Napoleonic power in Europe, but that did not mean that the regiment was unemployed. Far from it: in 1808 it was sent once again to the Caribbean to take part in operations against French possessions in the area. Once again, as had happened during its earlier deployment in the area, detachments were scattered throughout the area with presences on Barbados,

St Kitts and St Lucia. This was to be the regiment's home for the next nine years: it was not until June 1817 that it was ordered home and went into barracks at Northampton and Daventry. During the posting in the Caribbean time did not hang heavily on the regiment's hands as the men were in regular action against French forces. The battles might have lacked the historic importance of the fighting in the Peninsula, where Wellington was often outnumbered and had to make to do with the forces that he had, and not with numbers that he might have wished to possess, but as the regimental records make clear the fighting in the Caribbean was not without incident. For example, the 25th was involved in the capture of the French fort at Martinique on 2 February 1809 and with good reason the action was added to the regiment's battle honours:

> The Grenadier Company joined the Light Brigade, commanded by Major Campbell of the York Rangers. He, however, was wounded, when the command was given to Major Parsons of the 23rd Fusiliers, which regiment occupied a position in a wood near Fort Bourbon. This brigade, which consisted of the flank company of the 25th, the 63rd Regiment, and two black companies, crossed the country, and stormed a French picket under the walls of Beaulieu Redoubt. This picket, which consisted of one captain, two lieutenants and eighty privates, were all killed or wounded, with the exception of four or five sentinels who made their escape. Captain McDonald of the Grenadiers led his company to the gate, where he was challenged by the two French sentinels who were instantly shot. The brigade rushed into the place, and took the post under a heavy fire from Fort Bourbon; but as it was night the shot passed over the heads of the men, doing little or no damage. On the 19th of February the British batteries

> opened a heavy fire, which was kept up both day and night, on Fort Bourbon. On the 24th of February 1809, the fort, and all the dependencies of the island, surrendered.

Throughout the deployment there was a succession of casualties from death in action as well as from disease and the regiment required constant reinforcement. At the beginning of 1813, for example, 34 soldiers were drafted from the 16th Regiment (later The Bedfordshire Regiment) and the records show a similar pattern of reinforcement especially when other regiments were returning to Britain. The 25th's next action came a year later when it was involved in the successful capture of Guadeloupe at the beginning of 1810. For the most part, though, the regiment was allowed little time to stay in any one place and the records show that it was constantly on the move, changing station to meet the needs of the local situation. The records also show that for some of the men there were some narrow escapes:

> In the year of 1813, while Lieutenant-Colonel Light commanded the 1st Battalion of The 25th Regiment in the island of Guadeloupe, he dined one day with the Governor, and as he was riding home to the barracks (distant about one mile from the Governor's house) he was overtaken by a violent thunderstorm with heavy rain. A vivid flash of lightning coming very close to his horse, the animal took fright, and suddenly sprang over the precipice (which lay about five yards to the right of the road) of about thirty-four feet deep, into a river swollen considerably with the rain. The horse was killed by the fall, but Lieutenant-Colonel Light swam on shore with very little injury, and walked home to his barracks, a quarter of a mile distant from the place.

In addition to recording the life of the regiment during its tour of the Caribbean the records also reveal the many changes that took place in structure, personnel and uniform. At the end of 1810 cocked hats were replaced by caps for officers. In March 1811, at the request of the Prince Regent an allowance was provided for the purchase of wine for the Officers' Mess in lieu of the duty on wine allowed for officers of the Royal Navy. At the conclusion of the war against France in 1815 the 2nd battalion was disbanded at Cork and the remainder of the men were sent to Portsmouth for embarkation in order to join the 1st battalion in the West Indies. Infantry sword-exercises for officers and sergeants were introduced in 1817 shortly before the 25th ended its tour of duty with the thanks of the commander of the forces in the West Indies for the 'correct and creditable state of discipline evinced upon all occasions by this respectable corps during a service of many years in this command'. Once again the 25th was on its way home after a trying and successful deployment. There would be many more to follow in the months and years that lay ahead.

FOUR

Imperial Soldiering

Following the defeat of Napoleon at Waterloo it would be another 39 years before the army soldiered again in Europe. During that time the nation enjoyed four decades of peace and apart from the army's involvement in colonial campaigns it was a time of relative calm and prosperity. Inevitably it was also a time when the army suffered from the defence cutbacks that traditionally follow the cessation of hostilities; it was no different in the post-Napoleonic period that came later to be known as the 'long peace'. Although there were occasional 'invasion scares', most notably in 1846 and 1852, when there were fears of a cross-channel attack from France, the country's main shield continued to be the Royal Navy and the senior service managed to receive most of the available defence funding. Another consequence was the need to reinforce imperial defence. Fears that the United States might wish to indulge in territorial aggrandisement led to the creation of a 5,000-strong British garrison in Canada. That it was needed was proved in 1841–42, when there was a dispute over the Brunswick border and again in 1845–46 when the US and Canada almost came to blows over the Oregon border. Troops were also needed to put down disturbances in Demerara in 1823, Mauritius in 1832 and Ceylon in 1848. Fears that Maoris would attack European settlers in 1845 led to the garrison in New Zealand being increased to 15,000, troops were used in support of the civil power to maintain order in Australia and at home in Britain the 1840s saw an increase in political unrest, with the army being used to quell agitation by Chartists, early socialists who demanded universal male suffrage, the removal of the property qualification for membership of parliament and the re-drawing of electoral districts. Although the movement lacked a central organisation and was largely ineffective the government took fright at the huge demonstrations it inspired and responded with a heavy hand.

There was also a huge garrison in India, which was needed both to protect the holdings which had been gained in the previous century and to take part in the continuing expansion of the British Raj (British-controlled India). In addition to ill-fated attempts to bring Afghanistan under control in 1838–42 the period also saw the annexation of Sind in 1843 and the acquisition of the Punjab following the Sikh Wars of 1845–49. But India was not just warfare and campaigning. Throughout the nineteenth century and until India became independent in 1947 following Partition and the creation of Pakistan, the country was very much a home-from-home for the regiments of the British Army. Compared with service in the United Kingdom, life in India for a soldier was 'cushy'. Even the youngest or most recently enlisted private was treated as a 'sahib' but whatever their title, British soldiers were generally excused the kind of chores which would have been given to them at home in Britain. Cleaning up barracks was left to the sweepers, Indians did all the work in the cookhouse and the laundry was in the hands of the washer-women. In return a number of words entered the soldiers' vocabulary to be anglicised and used wherever a regiment was posted – buckshee (free, gratis), charpoy (bed), chit (written message), jeldi (hurry up), pukka (proper), tiffin (lunch or mid-day meal). Many are still in use and can be heard being used by British soldiers in the twenty-first century. Another innovation was the use of lighter clothes and by the 1880s khaki was in widespread use, although it did not become an integral part of army uniform until 20 years later. Apart from taking part in internal security duties or fighting the occasional war on the frontier the pattern of service for most soldiers in India was undemanding and mostly pleasant. Due to the excessive heat, especially during the dry season (April to October), all parades were over by mid-morning and there was no further activity until the cool of the evening. There was also a high premium on sport, especially for the officers, who were able to enjoy field sports and new and exotic activities such as polo. Then there was the country itself. As most British soldiers discovered, once experienced – India is a country which assaults all the senses, often simultaneously – men never forgot the country or its fantastic sights, sounds and smells.

For the 25th the end of the war against Napoleon meant a welcome return to a home station in 1817 and Ireland was its next destination. As happened to any infantry regiment based in that country the deployment meant that the companies were scattered around several towns, with the headquarters company at Templemore in Tipperary. Although this was a relatively quiet period in Ireland's history – an Act of Union had made the country part of the United Kingdom in 1801 – the calm was only on the surface. The clamouring for Catholic emancipation was growing in strength and would be encouraged by the creation of the Catholic Association in 1823 and the emergence of the radical lawyer Daniel O'Connell. In the year prior to the 25th's arrival there had been one of the periodic outbreaks of famine which brought death and misery to countless families. One visitor to the country, the novelist Sir Walter Scott, noted in his diary: 'Their poverty has not been exaggerated: it is on the verge of extreme human misery.' Faced by political grievance and the horrors of famine it was not surprising that the people of Ireland nursed grievances that would erupt into violent outbreaks of trouble throughout the century.

From the contemporary records it is clear that the regiment was frequently on the move in Ireland and to reflect its status as a home-based formation it was reduced in size from 800 rank and file to 650 rank and file with long-service men being discharged. Regular inspections show that the 25th was held in high regard, a general district order in 1821 noting that although the companies were scattered over several locations 'the same uniformity and good system was evident in the detachments as characterised that part of The Regiment stationed at headquarters'. The following year a similar inspection found that all officers and non-commissioned officers 'appeared perfectly to understand the execution of all duties required of them, and the precision of movements among the men, and their steadiness under arms, was most exemplary'. It is interesting to note that the orders frequently referred to the 25th as the Royal Borderers.

Towards the end of 1825 the regiment increased its establishment to 750 rank and file in ten companies in preparation for a return to the West Indies. The ports of departure were Cork and Kinsale and to reach them the scattered companies marched from Dublin through Kilkenny, Callan and Carlow, arriving at their destinations in the first week of November. The voyage to the Caribbean straddled the turn of the year and due to adverse winds the four service companies did not reach Barbados until the end of January 1826. Although the regiment was not on active service the tour of duty was not without its perils. Once again, as happened so often in that region, disease took its toll. In 1828 the death toll was one officer and 107 non-commissioned officers and men and the following year saw the deaths of 24 non-commissioned officers and men. Regular drafts from Ireland and Scotland kept the regiment up to strength but the haemorrhage showed no sign of abating – 31 officers and men in 1830, 58 non-commissioned officers and men in 1831, 27 rank and file in 1832, 20 non-commissioned officers and men in 1833, 21 non-commissioned officers and men in 1834 and 20 non-commissioned officers and men in 1835. It was with some relief that the regiment returned to Ireland in the late spring of 1835.

During the deployment in the West Indies there were further refinements to the regiment's uniform, arising from their earlier change of name and status. When new colours were being prepared in 1828 it was discovered that there was no authority for the regiment to use a white horse on its badge together with the words *In veritate religionis confido*. A memorandum from the War Office set the record straight and laid the ground rules for the badge, which was to be worn by the regiment and borne on its colours:

> It was therefore proposed that the badge of The Regiment should be 'The Castle of Edinburgh' with the motto *Nisi dominus frustra* [the motto of the city of Edinburgh since 1647] and the name of 'King's Own Borderers' placed round it. Also that the whole should be surmounted by a crown, to show that the corps is a Royal Regiment, and the new motto which His late Majesty George III commanded to be adopted should be placed in two corners of the colours attached to the crest of England, in contradistinction to the Lion of England which is borne by the 4th or King's Own Regiment. By this was attained the twofold objective of better identifying the motto *In veritate religionis confido* with the pleasure of the King, and also of more clearly explaining the meaning of it with reference to that of the motto of the city of Edinburgh.

In that way the regiment's origins in Scotland's capital were recognised but at the same time there was still some ambiguity as the royal crest was that of England. In 1961 this anomaly was investigated by the Lord Lyon King-of-Arms, who advised that the badge should be altered to include the royal crest of Scotland, a lion *sejant affronte erect*, that is, rising to prepare for action, imperially crowned and holding sword and sceptre, the crest which had been adopted by the Stuarts in the sixteenth century. Soundings were taken within the regimental family but the majority view was that the original royal crest should be kept even if it was not strictly Scottish. It was a good decision which reflected the regiment's chequered history; in time the lion came to be known, somewhat irreverently, as the 'Dog and Bonnet'.

On its return to Ireland the headquarters company was based once more at Templemore with companies at Cashel, Fethard, Bansha, Killeman, Thurles, Roscrea and Cappaghwhite. The strength of the 25th was fixed at 660 rank and file but this could only be maintained by regular recruiting as there was a constant drain of trained soldiers being transferred to other regiments on active service abroad. At the beginning of 1838, for example, the garrison in Canada had to be reinforced and the call for volunteers resulted in a steady stream of Borderers anxious to see service abroad and to accept the bounty of one guinea per man. At one point it looked as if the 25th was also bound for Canada but the orders were changed and instead of leaving Cork for Halifax the new destination was Cape Town in South Africa, where the regiment had been detailed to relieve the 72nd Highlanders (later 1st Seaforth Highlanders). In preparation the regiment sailed first to Devonport before departing for the long voyage south at the end of 1839. As the regimental records make clear, this was a complicated undertaking and gives some idea of the way in which the regiment readied itself for a move to the other side of the world:

> Two of the transports, the Pestonjee Bomanjee and the Lord Lyndoch, arrived in the [Plymouth] Sound about the middle of December. On the 21st five companies embarked as follows: the Headquarters, consisting of Lieutenant-Colonel Chambers, Captains O'Connor and Barnes (Acting Paymaster); Lieutenants McDonald, Pinder and Connolly; Ensigns Taylor, Northey, Wellesley; Quartermaster Potts and Surgeon Nivison; sixteen sergeants, nine corporals, seven drummers, two hundred and fifty-six privates, sixteen women and twenty-nine children on board the Pestonjee Bomanjee. Major D'Urban, Captains Jenkins and Guille, Lieutenants Gough and Lindsell, Ensigns Ogilvy and Smith, Assistant-Surgeon McDonald, eight sergeants, ten corporals, three drummers, two hundred and twenty-eight privates, seventeen women and twenty-five children on board the Lord Lyndoch. On Christmas Day, exactly fourteen years since The Regiment last sailed for the West Indies, the transports weighed, and dropped anchor again near the mouth of the harbour. On the 28th they stood out to sea and The Regiment commenced a fresh career of foreign service.

By the second week of January 1840 the ships had reached Tenerife but it was to be another two months before they finally anchored in Table Bay. The reason for the deployment to South Africa was occasioned by a long-running series of conflicts known at the time as the Kaffir Wars which raged for the greater part of the mid-century. This succession of conflicts with the Xhosa people, cattle-raising tribes of Eastern Natal, came about as the result of European expansionism as Dutch settlers began moving eastwards from the Cape in the 1770s. As the Dutch attempted to create settlements this led to fighting along the Great Fish River and they turned to the British in Cape Province for help. Although the use of force helped to settle the issue after a fifth war was fought in 1817, the British build-up led to increased strains not only with the Xhosa but also with the Dutch. In 1834 attacks on European settlements had forced the British governor, Sir Benjamin D'Urban, to drive the Xhosa back over the Great Fish River into a new settlement known as Queen Adelaide Province, where the Dutch were offered compensation for the loss of land. However, by then the Dutch and the British were unwilling to cooperate – incoming missionaries from Britain disliked the Dutch attitudes towards the native Africans – and so began the 'Great Trek' north which allowed the Dutch to create the new province of Natal and, in time, the Orange Free State. As we shall see, this arrangement did not solve matters but only created the problems which would lead the British and the Dutch to go to war later in the century.

The Xhosa forces were led by a tribal chief called Macomo and never numbered more than 7,000 but they proved to be spirited fighters, many of whom possessed firearms. During the course of the operations in Natal there were also heightened tensions with the Boers, who usually refused to obey instructions on the grounds that they were not British subjects. In 1842 this spilled over into confrontation when Boer farmers attacked a party of 250 soldiers from the 27th Foot (later The Royal Inniskilling Fusiliers) and prevented the survivors from getting away. To relieve them a detachment of the 25th under Major D'Urban was sent by sea to Port Natal in June and arrived to find that their landing was being opposed by the Boers, who refused all orders to surrender their position. Although the 25th succeeded in getting ashore and creating a bridgehead the official despatch written by Lieutenant-Colonel A. J. Cloete, the senior officer present, gives a good idea of the local conditions facing the detachments once they were on dry land. It also provides an accurate forecast of the Boers' fighting qualities and of the difficulties involved in engaging a mobile and committed force of citizen soldiers:

> The Boers abandoned their strong ground the moment we landed. Yet so thick was the bush and so broken the ground, that though, from the smart fire kept up, they must have been in force, yet not half-a-dozen of them were ever seen; and on the southern bluff, so thick was the wooded covering that nothing but smoke from the firelocks was seen. I have since learned that the number of Boers who defended the port amounted to three hundred and fifty men; their loss it has been impossible to ascertain . . . Captain Durnford was ordered to enter the bush on the right and drive the Boers before him, whilst I placed myself on a roadway in the centre, Major D'Urban taking the left along the harbour-beach.
>
> In this order we advanced through a bush the character of which it is difficult to describe, and which might have been held by a handful of resolute men against any assailants.

Not for the last time the Boers also showed that they understood the lie of the land, could read it well and were capable of using it to their own advantage – characteristics which they would rely on again later in the century while fighting the British. On this occasion though the men of the 27th were extricated but it was not the end of the incident. In the days that followed the Boers kept up their attacks on the British camp and to complicate matters further the Xhosa took the opportunity to renew their own assaults on the Boers and it took considerable military and diplomatic skill to prevent the incident spiralling out of control. The situation was eventually calmed down when the Boers under the leadership of Andreas Pretorius agreed to surrender to Cloete and place themselves under British authority. At the same time Natal was returned to British control.

Later in the year the regiment was spared further involvement in the fighting when it received orders to proceed to India, its first posting to the sub-continent. With them they took the grateful thanks of Sir George Napier, the commanding officer of the British garrison in South Africa, who made a point of praising the detachments of the 25th for their role in the recent action at Port Natal:

> The firmness, excellent discipline and gallantry displayed by that little band of British soldiers and their commander, under a constant and heavy fire from the insurgent Boers, as well as the several privations they endured for one month previous to their relief by the force under Lieutenant-Colonel Cloete is fresh proof of the indomitable courage and loyalty which ever animates the breast of the British soldier, and will show the rebel Boers the folly and hopelessness of their being able to withstand the power of the British government.

The regiment's next posting took them to Poonamallee near Madras with detachments at Arcot and Arnee. In January 1843 a number of men succumbed to cholera but throughout its stay in India the 25th received regular drafts both from Britain and from regiments returning at the end of their service. In 1846 129 volunteers from the 57th Foot (later 1st Middlesex Regiment) decided to throw in their lot with the 25th rather than return home, and at the beginning of 1847, while stationed at Cannanore, the regiment received 101 volunteers from the 63rd Foot (later 1st Manchester Regiment). The following year detachments from the regiment were ordered to Hong Kong to reinforce the garrison during a period of tension with the Chinese. This vital port and trading centre had been in British hands since 1842, when it was ceded by the Treaty of Nanking as an open port and in time it became one of Britain's most important colonial holdings in the Far East. On this occasion the trouble died down as quickly as it had flared up and the 25th did not proceed any further than Singapore, another vital port and trading centre. At the beginning of 1850 the regiment relieved the 51st Foot (later The King's Own Light Infantry Regiment) in the garrison at Bangalore and were soon in action restoring order at nearby Seringapatam, which was the scene of rioting between Hindus and Muslims. As was all too common, the trouble had been ignited by an incident when a pig's head had been thrown into the Golden Mosque. (In February 1792 Seringapatam had been the scene of a notable victory over Tipu Sultan during the Mysore Wars.)

In February 1855 the regiment received orders to return to Britain and men who wished to remain in India were given leave to transfer to the 43rd Foot (later 1st Oxfordshire Light Infantry) and 74th Highlanders (later 2nd Highland Light Infantry). By then the war in the Crimea had broken out (see below) but the 25th was not needed for the British force which had been sent to the region under the command of Field Marshal Lord Raglan. On returning to Britain the 25th moved north to Manchester, which was to be its home until July 1857, when it transferred to Dover. Its next station was Gibraltar following a period of home service which had only lasted two years and five months. On arrival the commanding officer, Lieutenant-Colonel Hamilton, was informed by the Deputy Adjutant-General that the 25th would be allowed to retain its pipers provided that 'the public is put to no expense for their clothing as pipers'. The letter also noted that the original permission for the pipers had been 'lost in time'. At about the same time it was agreed to form a 2nd battalion and this came into being at Preston in December 1859 under the command of Brevet Lieutenant-Colonel Allan, previously of the 81st Foot (later 2nd North Lancashire Regiment). At its embodiment it consisted of one lieutenant-colonel, two majors, 12 captains, 13 lieutenants, ten ensigns, one paymaster, one adjutant and quartermaster, one surgeon, one assistant surgeon, 53 sergeants, 49 corporals, 14 drummers and 932 rank and file. The records show that the majority of the officers came from other infantry regiments of the line while the newly joined ensigns are listed simply as 'gentlemen'. Following short periods at Aldershot and Shorncliffe the 2nd battalion moved to Edinburgh for the consecration of its first stand of colours in April 1863. An attractive description of the conditions inside Edinburgh Castle was written by a recruit from Sutherland which is quoted in Robert Woollcombe's history:

> They are very well off as to clothing and rations. They get two pairs of boots per year, one red coat, two pairs of trousers and light summer dress, and in fact everything they require, such as soap, black, whitening, shoe brushes, clothes brushes, shaving articles etc, etc, together with their rations and washing, all for eight and a half pence a day. Their rations are as good as I would wish to have. They have three meals per day, loaf and coffee at breakfast, potatoes, soup and beef at dinner and tea and loaf to supper. They have breakfast at 8 am, dinner at 1 pm and supper when they want it. They have not to cook their rations but get it cooked, only one of the mess men has to go and fetch it from the cookhouse. They have very good beds, every man has his own bed, which is about six feet long and three broad, the bedding comprises a mattress, a pillow, two coarse sheets and two blankets and bed cover.

Later that summer the 2nd battalion made ready for its first overseas deployment: on 28 July it sailed from Gosport to Ceylon on board the troopship *Himalaya*, taking its full complement of officers and men as well as 63 wives and 86 children. The final destination was not reached until the end of September and on arrival the battalion formed detachments at Trincomalee, Kandy, Galle and Nuwera Ellia, with the headquarters at Colombo. After four years of uneventful service – apart from 1866 when three officers and 15 soldiers succumbed to typhoid – the 2nd battalion was transported to Calcutta in January 1868. From there it marched to Shahjehanpur and later to Bareilly which was to be the battalion's home until November 1871, when it moved to Saugor. The route march took it through Cawnpore, Allahabad and Jubbulpore and only one man was lost – an unfortunate soldier who attempted to catch some ducks in a water tank and was drowned in the process.

For the 1st battalion there was a different set of circumstances. In June 1861 it moved from Gibraltar to Malta. This was to be its home until the summer of 1864, when it was unexpectedly transferred to Montreal in Canada to strengthen the British garrison at a time of tension with the US. The trouble had been fomented by American Fenian sympathisers who had announced their intention of invading Canada in support of their Irish cousins. (The Irish Republican Brotherhood, also known as Fenians, believed that Britain would never concede independence to Ireland without the use of physical force. The secret organisation had appeared the previous year and it had an influential input from expatriate Irishmen living in the United States, who provided money and weapons.) In the summer of 1866 they put their words into action when 1,000 of their number crossed the Niagara and entered Canadian territory and threatened other parts of the border. By way of response British Regular and Canadian militia troops were rushed into the area and 1/25th was taken by train and steamer to St John's on the River Richelieu. There it was joined by detachments of the Rifle Brigade and the force, under the command of Lieutenant-Colonel Fane, took up positions on rising ground called Pigeon Hill to counter the attempted invasion. It was a tricky position as detachments of the US Army were on the other side of the border, but following a desultory exchange of shots which resulted in the deaths of a handful of Fenians the danger quickly passed.

The end of the threat also ended the need to keep 1/25th in North America and on 3 August 1867 the battalion embarked on board the troopship *Tamar* for passage to Glasgow, where it based its headquarters, with detachments at Stirling, Ayr, Paisley and Dundee. A year later the battalion was on the move again, travelling by ship to Liverpool and then onwards by rail to Aldershot. Being at a home station meant a reduction in strength and the battalion's complement fell to three field officers, ten captains, 12 lieutenants, eight ensigns, five staff, 41 sergeants, 40 corporals, 21 drummers and 520 rank and file. All recruitment was stopped and in October 1869 the battalion moved again, this time to Anglesea Barracks in Portsmouth before crossing over to Ireland in December 1871 for a four-year tour of duty. It was the beginning of a time of momentous reforms within the British Army, and the 25th was part of the process.

In the second half of the nineteenth century the army underwent a number of radical changes to its regimental structure and the way it treated its soldiers. Some of these innovations were the result of technical developments such as the introduction of Snider and Martini-Henry rifles. (The latter was welcomed but not the former as it was found to be defective in the breech and required constant adjustments.) Other changes came about as the result of war itself. Two events stand out, and although neither involved the 25th, both were to have a substantial effect on the future of the army. The first was the Crimean War of 1854–56, which was fought to block Russian territorial ambitions in the Black Sea geo-strategic area and to prop up the Ottoman empire. In the late summer of 1854 a British–French campaign to reduce the Russian fortress and naval base at Sevastopol quickly ran into trouble and during the subsequent siege, which lasted until September 1855, British soldiers faced dreadful conditions during the harsh winter months. After the war it was revealed that only one in ten of the 19,584 casualties had been killed in action. The rest had succumbed to disease or the weather. As a result of the revelations of poor leadership and inadequate equipment, which had been made public in newspapers like *The Times*, changes were made to the operation and structure of the army, but given the prevailing conservatism many of the proposed reforms took time to take root. A Staff College came into being at Camberley to provide further intensive training for promising officers, the Crimean conflict having exposed the weakness of reliance on regimental soldiering alone. Recruitment problems were addressed by introducing short-service enlistment, the number of years being reduced from 21 years to six years with the Colours and six in the Reserves. As for the purchase of officers' commissions, which had been much criticised during the war, it was not abolished until 1871, the magazine *Punch* observing in a 'Notice to Gallant but Stupid Young Gentlemen' that they could only purchase their commission 'up to the 31st day of October. After that you will be driven to the cruel necessity of deserving them'. In fact the reform had little effect in regiments like the 25th, where the low rates of pay and the high cost of living meant that officers continued to come from the same social background as before – mostly from the upper and professional classes and the landed gentry.

The other great setback came hard on the heels of the Crimean War – the Indian or Sepoy Mutiny of 1857–59, which had a profound effect on the way the authorities viewed the administration of the sub-continent. In the summer of 1857 there was a serious outbreak of violence in India, involving Indian regiments of the East India Company's Bengal Army, which rapidly escalated to threaten the whole fabric of British rule. On 10 May 1857 the uprising began at Meerut, where the 11th and 20th Native Infantry and 3rd Cavalry regiments rose up against the local European population and started slaughtering them. The trouble had been simmering throughout the year and, amongst other grievances, the flashpoint was the decision to issue Indian troops with cartridges using the grease of pigs and cows, the first being unclean to the Muslims and the second considered sacred by the Hindus. The trouble spread at Cawnpore, where the garrison was slaughtered on 27 June despite promises of safe conduct, and at Lucknow, where the European population was besieged in the Residency by a force of 60,000 mutineers. Reinforcements from Britain were ordered to deploy immediately in India but the suppression of the disorder was not completed until two years later. Inevitably the incident caused a great deal of heartache on both sides. For the Indians it meant that governance of their country was removed from the East India Company, and it took time to recover the bonds of trust which had traditionally existed between the British and Indian soldiers. As for the British, they vowed that they would never again be placed in a position where their authority was in doubt. Prior to the outbreak of the mutiny there had been 232,000 Indian soldiers and only 45,000 British soldiers. After 1859 the ratio was changed to 190,000 Indian soldiers and 80,000 British soldiers, with British regiments being brigaded with Indian regiments.

There were also changes in the army's regimental system. For some time the War Office had toyed with the idea of introducing a territorial system, by which every regiment would be linked to its own local recruiting area. The result was the creation of Sub-District Brigade Depots which paired 141 infantry battalions at 69 brigade depots. Under this scheme, in 1873 the two battalions of the 25th were brigaded at York as 6th Brigade Depot, the regiment being deemed to belong to the English military establishment. However, further change was in the air. At the time all infantry regiments numbered 1st to 25th and the two Rifle Brigade regiments (60th and 95th) had multiple battalions, and plans were now prepared to provide all single-battalion regiments with two battalions through a process of amalgamation. Under a process begun in 1872 under the direction of the Secretary for War, Edward Cardwell, and finalised nine years later by his successor Hugh Childers, the remaining single-battalion regiments were to be linked with others of their kind to form new two-battalion regiments with territorial designations. Each regiment would also have local militia and volunteer rifle battalions consisting of part-time soldiers, the predecessors of the later Territorial Army

Driving the Cardwell/Childers reforms was the theory that one battalion would serve at home while the other was stationed abroad and received drafts and reliefs from the home-based battalion to keep it up to strength. As a result of the localisation changes regimental numbers were dropped and territorial names were adopted throughout the army but, as happens in every period of reform, the changes outraged older soldiers, who deplored the loss of cherished numbers and the introduction of what they held to be undignified territorial names, some of which bore no relation to the new regiment's traditions and customs. Being one of the

25 regiments which already had two or more battalions the 25th was not affected by the Cardwell/Childers reforms as there was no need for them to link up with another regiment, but far from continuing as before there was a radical change of name. Because the new brigade depot was situated in York the 25th became The York Regiment (King's Own Borderers), a move that effectively divorced the regiment from its historical origins in Scotland, not least with the city of Edinburgh.

Inevitably the change caused a great deal of resentment and the regiment complained bitterly about its loss of Scottish status. Initially the War Office was unsympathetic. The general view was that Scotland was over-represented in the British Army as it was and that recruiting would be a problem; the only compromise offered by the Adjutant-General was a change of location to Berwick-on-Tweed when the resident battalion, 5th Northumberland Fusiliers, moved to Newcastle. Eventually that sensible outcome was agreed and on 29 July 1881 the regiment moved to the depot at Berwick-on-Tweed, which would be its final home. The only drawback was that the Borderers would have no local militia or volunteer battalions and it was not until 1887 that the matter was finally settled. In that year the regiment's new title became The King's Own Scottish Borderers, with the depot at Berwick-on-Tweed and the headquarters of its 3rd Militia Battalion based at Dumfries. (Previously the formation had been the 3rd Militia Battalion of The Royal Scots Fusiliers but it had always been known as The Scottish Borderers Militia.) At the same time the regiment's recruiting area was designated as the counties of Dumfries-shire, Selkirk and Roxburgh, Berwickshire and Kirkcudbright, with Wigtonshire added later. The only changes were to the uniform: the red tunic of a line-infantry regiment was exchanged for a Highland doublet and trews of government tartan were now worn together with the Kilmarnock bonnet with dice-box border, introduced in 1864. As a result of the changes the 1st battalion became the overseas battalion and in 1875 moved to India where it was stationed at Fyzabad in the kingdom of Oudh. At the same time the 2nd battalion returned to Britain following a short tour of Aden.

FIVE

Afghans and Boers

During the period of the Cardwell/Childers reforms, as we have seen, the regiment had already started operating the home and overseas station system. In 1875 the 2nd battalion returned from India to become the home service battalion while the 1st battalion moved to India on board the troopship *Malabar*, landing at Bombay before moving to Fyzabad in Oudh. However, throughout the years before the century ended both battalions were destined to spend much of their time on active service in Afghanistan and India's North-West Frontier Province. The reasons for the British presence had come about as a result of the annexation of the Punjab in 1848 and had then been exacerbated by Russia's territorial ambitions in Afghanistan. As a result the frontier became an important, if awkward, area which required constant policing and that was never going to be an easy matter given the mettlesome independence displayed by the resident population. Seen on the map the frontier was defined by the River Indus and the plains and foothills which lay beyond it. And in the far distance towered the mountainous areas of the Hindu Kush, which were populated by the warlike hillmen of Waziristan. Not all of them were inclined to make trouble – the tribes in Baluchistan tended to be fairly pacific if left to their own devices – but it was a different matter dealing with the Pathan tribes in the area between Chitral and Baluchistan, where the British had to maintain control of the strategically important Khyber, Kurram and Bolan mountain passes. It was an area which was to become familiar to countless British soldiers who quickly learned that their opponents were hardy hillmen who, in the words of the poet Rudyard Kipling, did not give any mercy and did not expect to receive any:

> When you're wounded and left on Afghanistan's plains
> And the women come out to cut up what remains
> Just roll to your rifle and blow out your brains
> An' go to your Gawd like a soldier.

To meet the threat posed by the Pathan tribesmen the British Army evolved a policy which was known as 'butcher and bolt'. Basically this meant keeping its forces in the plains and only entering the mountainous tribal areas when there was trouble to be put down. When the tribes became too aggressive and started breaking the peace a punitive expedition would be mounted, the rebellious behaviour would be tamed by killing as many men as possible and then truces would be enforced which involved the surrender of weapons or the payment of a substantial fine. It was understood by both sides that no promises were binding and it was also recognised that violence would inevitably break out again after a suitable period of calm.

Later in the century the government in India attempted to develop a 'forward policy' of engagement with the tribal leaders to provide them with education and economic development in an attempt to break the cycle of violence. However, this was rarely successful because it was never developed wholeheartedly. Those in favour of the policy thought it ridiculous that large numbers of troops and resources should be tied down to solve a problem that was incapable of solution by force alone. Amongst the leading proponents was Lord Curzon, the forceful and energetic viceroy of India at the beginning of the twentieth century, who claimed that the problem could be resolved by the imposition of law and order and the subsequent implementation of education, economic aid and a modern infrastructure. It was a laudable hope – the same problem continues to exist in the country in the twenty-first century – but as Charles Chenevix Trench argued in his history of the regiments which guarded the frontier, 'there was among middle rank and junior officers a feeling . . . that if the Frontier was always at peace, India would be a much duller place; and that the army benefited immeasurably from annual war-games with the best umpire in the world, who never let a mistake go unpunished'.

Besides, the notion of a forward policy had not been successful in neighbouring Afghanistan, where Britain experienced mixed fortunes throughout the century. In one of the greatest military disasters ever to befall the British Army a brigade-sized force of some 4,500 British and Indian soldiers together with 12,000 camp followers was massacred while withdrawing from Kabul in Afghanistan in January 1842. The catastrophe was the climax of a policy to gain control of Afghanistan as a buffer to prevent Russian expansionism which might threaten Britain's holdings in India but the plans were badly thought-out and the whole operation was dogged by a lack of political will and insufficient funding. In the spring of 1839 a joint British and Indian force had crossed into Sind and marched up the Bolan Pass to take Kandahar, which fell without a battle. A British puppet ruler, Shah Shuja, was then placed on the throne and his rival, Dost Mohammed, was forced to flee.

It should have been the beginning of a period of settled rule but due to lack of money and muddled thinking in London the British garrison withdrew from Kabul at the end of 1841. This encouraged the Afghans to oppose Shah Shuja and the British quickly lost control in Kabul. A combination of atrocious winter weather, poor planning and Afghan duplicity led to the force's complete destruction. Despite Afghan promises of safe conduct through the passes to Peshawar the column was attacked and the only survivor was Dr William Brydon, an army surgeon, who managed to ride into the frontier fort at Jalalabad bearing the dreadful news. British opinion was outraged by the disaster and reinforcements were rushed out to India to lend assistance to the remaining Afghan garrisons at Jalalabad and Kalat-i-Ghilzai. In fact they would not be needed as Britain decided to abandon its Afghan policy and it was not until 1878 that there was further imperial involvement in the country's affairs. This is the point where The King's Own Scottish Borderers enter the story.

1st Battalion

The trouble in Afghanistan started during the 1860s, when a series of Russian annexations of neighbouring Tashkent, Samarkand and Khiva made the British believe that their rivals were attempting a pincer movement on the sub-continent. Events reached a climax in 1877 when the amir of Afghanistan, Shere Ali, entertained a Russian delegation but refused to allow a British mission to enter the country. His decision was a severe blow to national pride and the British decided that if Shere Ali would not allow the presence of a mission he would have one imposed on him. Three field forces were raised for the operation – the Kandahar Field Force commanded by Major-General Sir Donald Stewart, the Kurrum Valley Field Force commanded by Major-General Sir Frederick Sleigh Roberts VC and the Peshawar Valley Field Force under the command of

Lieutenant-General Sir Sam Browne, which contained 1/25th King's Own Borderers (its pre-1881 title). Before leaving for the front the men were forced to improvise khaki uniforms by having their white jackets dyed but it was a mixed success, with the results ranging from bright yellow to dark brown. As the railway only ran as far as Jhelum in the Punjab the battalion had to march 60 miles to Rawalpindi, where it was able to rest before marching the next 100 miles to Peshawar prior to the invasion of Afghanistan.

The jumping-off point for the operations was Jamrud, where the battalion War Diary noted tersely that the intention was 'to impress the inhabitants of these regions with an idea of British Power and the feasibility of entering their territory when desired'. In practice this meant taking command of the high land and creating strong points of stone sangars – reinforced piquet points which provided the main defence against attack. While 1/25th was engaged on operations in the Bazar Valley Roberts's field force encountered the enemy blocking the Peiwar Kotal pass in a seemingly impregnable position but his battalions succeeded in outflanking and beating off the Afghans. The road to Kabul was now open. Shere Ali made good his escape – he died in February the following year, leaving the British free to enter into a new treaty which was signed at Gandamak with his son, the new amir, Yakub Khan. This lull in the hostilities allowed Browne's field force to retire at the end of May and the battalion marched back to Peshawar, where it faced high temperatures and an outbreak of cholera.

By the time spring came to the country all appeared to be quiet and to the British it seemed that their tactics had worked. Under the terms of the treaty Yakub Khan was granted a subsidy in return for accepting a new British mission led by Pierre Louis Cavagnari, a diplomat whose father had served as a soldier under Napoleon. However, Cavagnari misjudged the situation by failing to appreciate the strength of the anti-Western sentiments within the country and he paid for that misunderstanding with his life. In the middle of September news reached Simla that Cavagnari and his entire embassy had been murdered. As the other field forces had retired back to India Roberts had the only available troops and he marched them rapidly to Kabul in order to carry out the orders he had received from Calcutta 'to strike terror and strike it swiftly and deeply'. (This appealed to Roberts, who wrote later in his memoirs that he would not have inserted Cavagnari's embassy until 'we had instilled that awe of us into the Afghan nation which would have been the only reliable guarantee for the safety of our Mission'.) At the beginning of 1880 1/25th joined the 2nd Division of Roberts's field force and took part in operations between Landi Kotal and Fort Dakka in the Jalalabad Plain. Following similar punitive operations in the Laghman Valley the 2nd Division was renamed the Khyber Line Force, charged with the responsibility of guarding the lines of communication between Kabul and Peshawar.

Kabul had fallen to Roberts's force but the *jihad*, or holy war, continued. In July the following year a British force was annihilated at Maiwand in southern Afghanistan by an Afghan army led by Ayub Khan, Yakub Khan's brother, and the news sent shock-waves through the garrison when it reached Kabul. Out of 2,476 men, 934 had been killed and 175 were wounded. Then came the news that another force under Major-General Sir James Primrose was besieged in Kandahar. Roberts wasted no time in drawing up a relief force of 10,000 men plus beasts of burden and camp followers which took 23 days to cover the 350 miles over trackless country in trying physical conditions. On 31 August 1880 the column reached Kandahar to find the garrison so dispirited that they lacked the will to fly the Union flag. Around the city the Afghans were positioned in the high hills and to Roberts's infantry fell the responsibility of flushing them out. These last skirmishes ended the Second Afghan War and, having put Abdul Rahman, a nephew of Shere Ali, on the throne, the British were able to withdraw once again from the country.

But for personality problems in the command structure 1/25th would have taken part in the operations to relieve Kandahar. Following the loss of the commanding officer, who had been invalided home during the recent cholera outbreak, the battalion was commanded by Major F. S. Terry but he found himself in constant disagreement with the other senior officer, Major N. C. Ramsay. According to the reminiscences of Lieutenant (later Lieutenant-General Sir) Charles Woollcombe, the two officers 'had quarrelled since they were subalterns and they could not serve together'. Woollcombe also noted that Terry 'had a way of fighting with the General and Staff' and from his subsequent career he seems to have been an idiosyncratic officer. After retiring from the army in 1881 he served as a trooper in the Bechuanaland expedition of 1885 and attempted without success to offer his services during the Boer War (1899–1902) and the First World War (1914–18). A great son of the regiment – his father had been commissioned in the 25th in 1799 and had fought at Egmont-op-Zee – Major Terry died in 1933, aged 94.

At the conclusion of the operations in Afghanistan 1/25th moved to the hot-weather station of Cherat where it became the 1st battalion of the York Regiment, King's Own Borderers, its title until 1887 when it became 1st King's Own Scottish Borderers. (For ease of recognition, it will now be referred to as 1st KOSB.) For the next few years the battalion's life was to be dominated by route marches in the Punjab. Each day's march began at seven o'clock in the morning and continued to about eleven o'clock. During this period the men of 1st KOSB saw most of the main towns of the area – Rawalpindi, Jhelum, Ludhiana, Ambala, Simla and Dagshai, a roll-call of the British presence in the Punjab in the heyday of Queen Victoria's empire in India. This pattern continued until

October 1899, when the battalion returned to the plains and made its way to Calcutta prior to a new deployment in Burma, where it was engaged on internal security duties in the northern part of the country.

These operations were to engage the battalion for the next 13 months and they came about as a result of the East India Company's decision to invade Burma in 1825 by sending forces to invest the capital, Ava, by way of advancing up the Irrawaddy. As a result Britain annexed the coastal area centred on Chittagong and the Burmese agreed to stop interfering in Assam, the reason for the war in the first place. Further wars had taken place in 1852–53 and 1885–87 but the northern areas were never fully pacified, hence the presence of 1st KOSB in what became known as the Chin Lushai Expedition. Operating in two columns, British forces moved into the Chin Hills between the Assam border and the River Chindwin to punish local tribes who were attacking over the border into India. This proved to be bruising work carried out in a hostile jungle environment and with few supplies owing to the long lines of communication. A further problem was the elusiveness of the enemy. The official history of the Third Burmese War noted that the Chins usually refused battle and, far from accepting the British presence, 'they all seem to look upon us as being in their hills more on sufferance than by right of superior force'. The main casualties in 1st KOSB were caused by sickness and disease – 21 dead and 279 invalided out of the operational area. It was not until December 1890 that the battalion made its way to Rangoon and a return to Britain.

For the next nine years 1st KOSB was the home service battalion, serving mainly in southern England but also taking part in regular recruiting drives in the Scottish Borders, its main source of new soldiers. As the century came to an end growing tensions in southern Africa plunged Britain into a costly and humiliating conflict with the Boers, Dutch immigrants who had settled in Cape Colony. Co-existence proved to be impossible and the Boers had started moving out in the first half of the century and had trekked north to establish Transvaal and the Orange Free State. However, that migration did not solve matters and, as we have seen, the enmity had erupted into open war in 1880. Following a humiliating defeat at Majuba Hill – where a lightly armed Boer force outfought and outwitted the British – an uneasy peace had been restored, with the Boers operating self-government under British suzerainty. It was a compromise and sooner or later it was bound to be tested. The flashpoint came in 1886 with the discovery of seemingly limitless supplies of gold in Boer territory south of Pretoria. Lured by the prospect of untold riches speculators flooded in from Britain and Europe and before long the Boers were outnumbered by outsiders who appeared to be threatening their traditional conservative way of life. To prevent that happening and to protect the interests of his fellow Boers in the Transvaal, President Kruger passed stringent laws excluding non-Boers from participation in political life while retaining the right to tax them.

Such a state of affairs was bound to cause trouble but when it came in 1895 it proved to be a botched business. Acting in the mistaken belief that an uprising against the Boers was imminent the British imperial adventurer Cecil Rhodes encouraged his associate, Dr Starr Jameson, to lead a raid into the Transvaal to bring down Kruger's government. It was an abject failure, but the Jameson Raid had far-reaching consequences. Rhodes was disgraced, Britain was made a laughing stock and to make matters worse the subsequent negotiations to retrieve the situation settled nothing. Each new concession was met with further demands and gradually war became inevitable. In 1899 Britain despatched 10,000 troops to South Africa to bolster her garrison while the Transvaal, now backed by the Orange Free State, made plans for mobilisation. War was declared on 12 October after Kruger's demands that Britain remove her troops from the frontier were ignored in London and within a week General Sir Redvers Buller VC was on his way to South Africa to take command of the imperial forces in what everyone hoped would be a short sharp war. It proved to be nothing of the sort and ahead lay an embarrassing succession of defeats which fully tested the will and determination of the British Army.

At the beginning of January 1900 1st KOSB left for South Africa on board the liner SS *Braemar Castle*, 'not a particularly fast boat but very comfortable and roomy' according to one officer. Following the hasty recall of reservists at Berwick-on-Tweed the battalion's strength was 28 officers and 1,082 other ranks. At the same time the 3rd Militia Battalion was embodied and moved to Belfast prior to embarkation for South Africa. Its task was to guard the lines of communication and these part-time soldiers were to give a good account of themselves together with the militia reservists of other line-infantry regiments. All told, 45,000 militia soldiers saw war service in South Africa. Militiamen had always served in the British Army as a defence force, originally selected by ballot, but the part-time soldiers had often proved to be a mixed blessing. In 1852 a new act put the force on a sounder footing. While recruits continued to be volunteers they could transfer to the Regular army after three months' training or decide to stay on as part-time soldiers provided that they agreed to train for a month in every year. To give coherence to the scheme, under the Cardwell/Childers reforms militia battalions were linked to the Regular battalions of their parent regiment. According to the records of 2nd Lieutenant W. F. Graham (quoted in Woollcombe's regimental history) the KOSB militia men were not inexperienced amateurs but 'nearly all old soldiers . . . and about as tough and drunken a lot as one could wish to see'.

Both battalions arrived in South Africa at a particularly difficult and demoralising time. In the second week of December 1899 the Highland Brigade commanded by Major-General Andrew Wauchope had suffered a humiliating defeat at Magersfontein in which the casualties were 210 killed and 738 wounded. The misery was compounded by news of two other heavy defeats at Stormberg and Colenso, a period of setbacks that the war correspondent Arthur Conan Doyle christened 'Black Week'; morale was at an all-time low. In an attempt to retrieve the situation the troop levels in South Africa were increased, Buller was sacked and command of the army was given to Lord Roberts of Afghanistan fame, who had won the Victoria Cross during the Indian Mutiny. His chief of staff and second-in-command was General Sir Horatio Herbert Kitchener, fresh from his exploits in subjugating Islamic fundamentalists in Sudan (see below).

On arrival in South Africa on 26 January 1900, 1st KOSB was brigaded with 2nd Lincolns, 2nd Hampshires and 2nd Norfolks in 7th Division and moved immediately to the Orange River where Kitchener was in the process of bludgeoning the Boers into submission at Paardeberg. The battalion War Diary recorded that the march to the battle-front had taken 19 hours and that the men had covered 26 miles only to arrive 'in an utterly exhausted condition which was accentuated by the absence of food, the baggage with rations being far behind'.

Paardeberg was a hard-pounding battle which Kitchener knew he dared not lose, and his tactics reflected that necessity. He made use of his superior fire-power and this time his infantry showed greater resolve than they had shown in the previous year. A diary kept by Private William Fessey, one of the battalion's Maxim gunners, provides a vivid picture of the dynamics of combat during the battle:

> We let them come within 300 yards of us and then we started, and rattled into them with Maxim and [rifle] volleys for all we were worth. We emptied belt after belt and I could see them fall and I am sure they suffered heavily. We could see their horses fall by scores and running about riderless, they got a surprise. They made a rush for a kopje where one of our companies were, and the Boers were driven back with heavy losses three times that morning [23 February]. We was lucky on our side, we only had five men wounded and one officer.

Four days later, at three o'clock in the morning, 1st KOSB was stood to arms but instead of preparing to fight another battle they saw white flags flying along the Boer lines and later in the day the Boer leader Piet Cronje surrendered his forces. Fortuitously it was also the anniversary of the earlier British setback at Majuba Hill. This was the turn of the tide and the war entered a new phase with the invasion of the Orange Free State and the Transvaal. By September Pretoria and Johannesburg had been occupied and the Boer army under the Commandant-General Martinus Prinsloo had capitulated at Brandwater Basin. During the latter phases of the war much use was made of a new force of 20,000 Mounted Infantry (MI) troopers. These were ordered by the War Office as 'a matter of immediate urgency and permanent importance' and its men were supposed to 'shoot as well as possible and ride decently'. Acting as scouts and rapid response forces, they were to be one of the more successful innovations of the war. The majority came from dominion forces and from British yeomanry regiments but at least a quarter of the number were raised from Regular battalions, most of whom were reservists with experience of horses. MI troopers from 1st and 3rd KOSB were posted to the 7th Mounted Infantry Regiment and as the diary of Lieutenant Chandos Leigh makes clear, in the early days at least, things did not always go according to plan:

> I shall never forget the day we took over our mounts. In the first place not one of the officers had any previous Mounted Infantry experience, and the few of our men who had were men who had served many years ago in one MI section in Egypt. The rest were chosen from men who were supposed to have a knowledge of horses, viz, any man whether he had been an officer's groom, driven a milk cart or a hansom cab or even sold cats' meat, would according to himself be eligible. Therefore, when we proceeded to draw our cobs, all fat underbred looking Argentines, there was chaos. Every man went for the cob he liked the look of best, these beasts being all more or less wild resented being bridled and still more being backed. In five minutes the whole place was a mass of loose ponies flying about with bridles round their legs, and Tommies vainly endeavouring to mount bareback and being put off. I was shot off myself the moment I tried to mount.

However, from those unpromising beginnings the Mounted Infantry was one of the successes to come out of the Boer War and the regiment's first Victoria Cross was won by Lieutenant G. H. B. Coulson, adjutant of the 7th Mounted Infantry Regiment (see Appendix). All told, The King's Own Scottish Borderers lost 97 men in the Boer War, which finally came to an end with the signing of the Treaty of Vereeniging in May 1902. On the 1st battalion's return from South Africa permission was given for The King's Own Scottish Borderers to wear trews of Leslie tartan in honour of their founder. Ireland was to be the battalion's home until 1905, when it moved to Colchester prior to a new deployment in Egypt and Sudan. In 1911 the battalion embarked at Port Sudan for India, where it was to be resident for the next three years.

2nd Battalion

At the time of the final stages of Cardwell/Childers reforms the 2/25th had been based in Ireland at Fermoy, Kinsale and Dublin. In 1887, when it became 2nd KOSB, it moved to Aldershot where it took part in the Golden Jubilee celebrations to commemorate Queen Victoria's 50 years on the throne. However, the battalion was soon on the move again in the following year when it was ordered to deploy to Egypt to reinforce the British garrison at a time of increased unrest by Islamic fundamentalists led by Mohammed Ibn Al-Sayd Abdullah. A teacher from Dongola province, he had proclaimed himself the Mahdi, the 'expected one', descendant of the prophet Mohammed, who intended to rid Sudan of infidel forces. To oversee the necessary withdrawal from the area under the Mahdi's control the British government despatched General Charles Gordon to Khartoum, where he and his staff were quickly besieged. A relief force set out to save him but it arrived too late – Gordon was murdered on 26 January 1885 – and the region was set for a further 13 years of muddle and unrest. There was a deterioration in the situation in Sudan following the creation of an Islamic fundamentalist state overseen by the Mahdi's successor, Abdullah Ibn Muhammed, known as the Khalifa. During his despotic rule those who offended him were subjected to mass murder, rape and savagery, and by the 1890s there were calls for him to be deposed and for Sudan to be brought under British–Egyptian rule.

On arriving in the area of operations 2nd KOSB joined the Suakim Field Force under the command of Major-General Sir Francis Grenfell and immediately deployed to the Red Sea port of Suakim in December 1888. The battalion was soon in action with the dervishes, as the Khalifa's followers were called. Still wearing their red tunics the men of 2nd KOSB were conspicuous and offered easy targets as they constructed a defensive position out of thorn bushes known as a zariba. Happily the dervishes had removed the backsights from their weapons in the mistaken belief that they were a hindrance and the battalion suffered few casualties as a result. After the Battle of Gemaiza (the name of the southern defensive position) Grenfell reported that never before had he had 'a more handy, intelligent battalion' under his command. Despite the setback at Gemaiza the Khalifa pushed ahead with plans to invade Egypt and was eventually defeated decisively at the Battle of Toski in August 1889.

The following year 2nd KOSB ended its tour of duty and embarked at Suez for India. When it arrived in Bombay its strength was 25 officers, two warrant officers, 28 sergeants, 14 drummers, 37 corporals, 750 privates, three officers' wives and five children, 22 soldiers' wives and 36 children. From Bombay the battalion moved to Sabathu near Simla but it was soon to be in action again when it left for the mountainous reaches of the north-west, where another rebellion was in the offing. This time the unrest flared in Chitral, which lies between Kashmir and Afghanistan, and once again the trouble was caused by tribal jealousies and animosities, and the murder of a native ruler. When the British Political Agent was besieged in Chitral with a small garrison of Indian troops the government decided to act. A relief force was put together under the command of Major-General Sir R. C. Low and the battalion served with 1st Gordons in 2 Brigade under Brigadier-General H. G. Waterfield. In wet and cold weather conditions the force set out through the Swat Valley for the Malakand Pass, where 2nd KOSB went into the attack with 1st Gordons. In the vanguard were the Guides Infantry and 4th Sikhs, who received fire support from the accompanying mountain batteries and Gatling guns but the final attack still demanded a frontal assault. This was entrusted to the Borderers in the centre with the Gordons on the right and both battalions were soon in contact with the enemy above them. Their fortitude under fire prompted the soldier and explorer Sir George Younghusband to write a spirited account of the action in his later history of the campaign:

> It was a fine and stirring sight to see the splendid dash with which the two Scottish regiments took the hill. From the valley to the crest at this point the height varies from 1,000 to 1,500 feet and the slope looks for the most part almost perpendicular. It was this very steepness which partly accounted for the comparatively small loss suffered from the enemy's fire and the shower of huge boulders which were hurled upon the assailants; but the chief reason for this happy immunity was the wonderfully spirited manner in which the men rushed breastwork after breastwork and arrived just beneath the final ridge before the enemy had time to realise that the assaulting columns were at their very feet.

During the action 2nd KOSB lost two men killed and 14 wounded but after five hours of intensive fighting the Malakand Pass had been secured. Ahead lay the upper reaches of the Swat Valley with some difficult river crossings but the show of strength discomfited the rebels and forced them to give up their siege of Chitral at the end of April. The campaign was over and 2nd KOSB had added to the regiment's reputation, not least because the men remained focused not just on the task in hand but also on maintaining discipline. The battalion War Diary claimed that it was 'the most healthy regiment on the expedition' largely because, as one soldier claimed, it offered a good opportunity for those who wanted to abstain from alcohol and become teetotal. In August the battalion returned to Rawalpindi but it was soon to be in action again in another punitive expedition against the warlike tribesmen of the north-west. This was with the Tirah Expeditionary Force under the command of Major-General Sir William Lockhart, which was formed to put down a rebellion by the Afridi and Orakzai tribes which had been attacking British positions along the frontier.

Between them the two tribes raised an army some 50,000-strong and quickly began causing severe mayhem in the Khyber and Kohat areas, attacking outposts and murdering anyone who got in their way. This time 2nd KOSB served in the 2nd Division and took part in the main battle of the operation at the Dargai Heights in October 1897. Although most of the plaudits went to 1st Gordons, whose Piper Findlater was awarded the Victoria Cross for continuing to play his pipes although wounded in both legs later in the battle, 2nd KOSB also took part in a frontal attack on the enemy positions in the early stages. Attacking in tandem with 1st/3rd Gurkha Rifles, the battalion had to move across rocky open ground to get to its objective, and all this had to be achieved through what Lionel James, an accompanying Reuters correspondent, called 'sheer climbing power'. A failure to consolidate meant that the ridge had to be retaken two days later but the defeat of the Orakzai tribesmen dissuaded them from engaging in further set-piece battles. While this came as a relief the tribesmen continued to attack and harry Lockhart's men as they pushed into Tirah with over 40,000 transport animals and assorted camp followers. Some idea of the difficulties they faced can be found in James's later history of the campaign:

> General Lockhart was called upon to take this undisciplined multitude across mountains where there was not even a vestige of a goat-track, along river-beds strewn with boulders where the only path was knee-deep in icy water,

> through gorges where two animals abreast closed the passage, up gradients which in their initial stages defeated even mountain battery mules, down descents which were almost precipices; through barren places where food there was none, and into extremes of climate which destroyed the weaklings and consequently increased the loads of the more robust.

The tribesmen added to the problems with hidden rifle fire on the British columns and most of the force's casualties came from snipers. With winter approaching Lockhart withdrew his army to the Peshawar Valley and during the withdrawal the men came under unremitting attack as they marched out of the Tirah area. During the march back to Swaikot the battalion lost eight men killed and 43 wounded. From there 2nd KOSB made its way to Cawnpore, where it received its trews in Leslie tartan in April 1900. Ahead lay a deployment at Maymo in Burma in 1903 and a short stay in Aden in 1906 before the battalion's return to Glasgow, where it took up residence in Maryhill Barracks. This was to be the battalion's home until 1910, when it made the short voyage across St George's Channel to be one of the resident battalions in Northern Ireland, based at Palace Barracks, Holywood. This was a tense time in Irish affairs. Nationalist sentiments were on the increase, with calls for independence being channelled through groups like Sinn Fein, the Gaelic League and the Irish Republican Brotherhood. A home rule bill was due to be passed in 1912 which would have provided a parliament in Dublin, albeit with limited powers, but this was opposed by unionists in the six Ulster counties of Down, Derry, Antrim, Tyrone, northern Fermanagh and north and mid-Armagh.

To reinforce their beliefs both sides had started arming themselves, mainly from Germany, and the Ulster Volunteers, raised by Sir Edward Carson had 100,000 members who began drilling in the Ulster counties, the organisation being provided mainly by the Orange lodges. As tensions began rising and it became clear that the Ulster counties would resist any move which gave Ireland home rule, if necessary by using force, the tensions spread into the British Army's garrison in the country. Inevitably this affected 2nd KOSB, which had moved south to Royal Barracks in Dublin. Although the army is supposed to be apolitical there was a good deal of sympathy for the Ulster Protestants and the question began to be posed: would the Irish-based regiments act against the Ulster Volunteers in the event of any outbreak of violence? Matters came to a head in March 1914 when the War Office let it be known that any officer with family connections in the Ulster counties would be allowed to 'disappear' temporarily but that others who disobeyed orders would have to resign or be dismissed from the service. When the question was put to 2nd KOSB all the officers bar the commanding officer, the second-in-command and the adjutant chose resignation. The matter escalated when 59 officers serving with the 3rd Cavalry Brigade at the Curragh garrison outside Dublin also chose resignation.

As a result of the incident, which came to be known as the Curragh Mutiny, Lieutenant-General Sir Charles Fergusson, commanding 5th Division, appealed to the officers to rethink their position and his good sense helped to still passions. Ten officers in the battalion had decided to hold out but Fergusson's intervention made them change their minds. However, the incident had revealed the strength of Ulster's unionist beliefs and was to have lasting repercussions in the country. It was also not the end of the regiment's involvement in Ireland's affairs. In the aftermath of the Curragh Mutiny gun-running increased, with both sides importing huge numbers of rifles from Germany. While the Ulster Volunteers acted with impunity, steps were taken to prevent weapons being imported into the south and this led to confrontations with the security forces. The most serious incident came in the summer of 1914 when the nationalist Erskine Childers arrived at Howth in his yacht, the *Asgard*, bringing with him 900 rifles and ammunition. On 26 July the battalion diary recorded:

> When most of the officers were out of the barracks, a telephone message arrived at about 3.30 pm to the effect that National Volunteers were landing rifles near Howth and that 100 men of the battalion were to proceed by tram at once to support the Police, taking 100 rounds a man and using force if necessary.

Despite police intervention the weapons were unloaded in half an hour and by the time the KOSB detachment arrived at Clontarf the volunteers had dispersed, leaving a small crowd to jeer at the police and the soldiers. As the detachment made its way back to Royal Barracks there were scuffles and outbreaks of stone-throwing which continued over the three-mile march back into Dublin. Tensions were high when the soldiers turned off O'Connell Street onto one of the quays by the River Liffey, known as Bachelor's Walk, and as the violence increased the rearguard wheeled to face the crowd. An officer who had just arrived gave the order for the crowd to disperse and in the confusion the men opened fire. It is not clear if the order was actually given or if by raising his hand for silence it was thought to have been given, but a woman and two men were killed, even though no soldier had fired more than two rounds. In the subsequent inquiry the government's commission concluded that the soldiers' action had been 'tainted with illegality' and the battalion diary subsequently recorded the outcome for the KOSB: 'The feeling against us in Dublin was very bitter and the battalion was strictly confined to their barracks.'

Coming on top of the Curragh Mutiny the Bachelor's Walk shootings reinforced an idea in nationalist minds that the government favoured the unionists but the resultant storm was soon to be overshadowed by a greater tempest taking place in Europe. Even so, the incident left a long shadow over the regiment and for many years nationalists in Ireland would refer to it as the King's Own Scottish Butchers.

SIX

The First World War: The Western Front and Italy

The summer of 1914 dawned as it usually does in Scotland – with thoroughly unpredictable weather. July and August, the traditional holiday months, are often wet and windy with few days of prolonged sunshine and the fourteenth year of the new century was no exception. Later, people would remember those summer months of 1914 as a long and pleasurable interlude before the onset of war, the last golden age before the world was plunged into the horrors of global conflict, but the weather records of the period show a somewhat different picture, as far as Scotland was concerned. While there were days of sunshine, the polar opposite also prevailed: uninterrupted summer skies were interspersed with longer periods of rain and blustery weather when grey skies were universal. In the first half of the month King George V and his wife Queen Mary made a royal visit to Scotland and they did not escape the vagaries of the weather. It rained in Edinburgh for much of the visit but when the King and Queen were in Lanarkshire on 12 July the *Scotsman* reported 'beautiful weather' and insisted that the visit had been a great success, underlining the importance of the monarchy to the country and its people. Scotland's summer months are often plagued by cold spells and showers but, even so, holidaymakers in 1914 were apparently not put off by the prospect of unseasonable weather. The *Glasgow Herald* reported that record crowds had flocked to fashionable east-coast beach resorts such as North Berwick and Elie and the city's trades fair fortnight had seen large numbers of day-trippers and holidaymakers taking the traditional voyage 'doon the watter' from the Broomielaw in Glasgow to the resorts on the Clyde estuary. Special trains were also run to resorts in England, Blackpool being a popular destination, while other holidaymakers took steamers to the Isle of Man and further afield to resorts in Northern Ireland.

One other group of citizens was also enjoying the annual break from the stresses and strains of work – the part-time soldiers of the Territorial Force who were engaged in their annual two-week camp as part of their commitment to training. Formed in 1908 for home defence, the Territorial Force built on the former militia and rifle volunteer system and consisted of volunteer battalions which were attached to the Regular infantry regiments, and their role was home defence in time of national emergency. Being a part-time soldier offered other benefits: it was companionable, offered self-respect and produced steadiness of character, all considered to be moral virtues in the Scotland of the period. In The King's Own Scottish Borderers there were two Territorial battalions – 4th (Border) Battalion and 5th (Dumfries and Galloway) Battalion – and both formed second and third line battalions for training and reinforcement. They were well supported too: in the 1/5th battalion the towns of Whithorn and Kirkconnel supplied more volunteers per head of population than any other part of Britain. Typical of the spirit which imbued the volunteer soldiers was the example of Professor George

Francis Scott Elliot, a noted botanist, traveller and author from an eminent Dumfries-shire family who re-enlisted, although over-age, and went on to write the war history of the 1/5th battalion. To give some idea of the latter battalion's geographical spread, the regimental records show the location of the eight (later decreased to four) rifle companies:

A Company – Dumfries (detachment at Moniaive)

B Company – Annan (detachments at Langholm and Canonbie)

C Company – Lockerbie (detachments at Ecclefechan and Moffat)

D Company – Sanquhar (detachments at Thornhill and Kirkconnel)

E Company – Maxwelltown

F Company – Dalbeattie

G Company – Castle Douglas (detachments at Corsock, Gatehouse and Kirkcudbright)

H Company – Newton Stewart (detachments at Wigtown, Creetown, Kirkcowan, Whithorn and Garlieston)

In normal times the summer camp was an opportunity for the battalion to train together, to engage in shooting practice and to sharpen up their military skills. For many men it was also their only chance to get away for a break and for the poorer it was an opportunity to enjoy a holiday with pay (one shilling a day), an important consideration at a time when paid holidays were not always the norm. But 1914 was different. No sooner had Territorial battalions all over the country finished their annual camps than they found themselves on a war footing. As the summer unfolded Europe found itself sliding towards war as a result of an assassination which had been barely reported in the British press.

In June, while visiting Sarajevo, the capital of Bosnia-Herzogovina, the Archduke Franz Ferdinand, heir to the throne of Austria-Hungary, had been gunned down together with his wife. When it became clear that neighbouring Serbia might have been implicated in the attack the crisis deepened. On 23 July, weeks after the assassination, Austria-Hungary issued an ultimatum to Serbia, making ten demands for the suppression of Serb nationalist groups, the punishment of the assassins and participation in the judicial process. Serbia was given 48 hours to comply and although the response was placatory its government stopped short of allowing Austria-Hungary to take part in the trial of the assassins, arguing that the matter should be referred to the International Court at The Hague. That readiness to cooperate seemed to settle the matter, but already diplomacy was proving powerless to stop Europe's drift towards war. Both countries mobilised their armed forces when Germany, Austria-Hungary's main ally, encouraged Vienna to take decisive action against the Serbs before any other country intervened in the crisis. Confident of German support, Austria-Hungary declared war on Serbia five days after issuing its first demands, thus paving the way for a wider conflict. The following day, 29 July, Russia, Serbia's traditional friend and protector, began to deploy its forces along the border with Austria and within 24 hours this was followed by the order for full mobilisation.

Although the move was made to discourage Austria it threatened Germany, which immediately demanded that Russia 'cease every war measure against us and Austria-Hungary'. On 1 August Germany declared war on Russia, followed two days later by a further declaration of war against France, Russia's ally. That same day German forces began crossing into Belgium as part of a pre-arranged plan to encircle Paris from the north. Britain, which had wanted to remain aloof from the crisis and was not formally in alliance with any of the main participants, was now about to be pressed into the conflict through a treaty of 1839 which guaranteed Belgium's neutrality. On 4 August, no answer having been received to an ultimatum that Belgium should remain unmolested, Britain declared war on Germany. As part of the 'Precautionary Period' of the Defence Plan Prior to Mobilisation, formations of the Regular Army based in Britain were told to return to their depots on 29 July. Most were on their annual summer camps or undertaking live firing exercises. As we have seen, at the outbreak of war 2nd KOSB was based in Dublin, where it formed part of 13 Brigade in the 5th Division, while the 1st battalion was based in Cairo. The British-based battalions crossed over to France with the first elements of the British Expeditionary Force (BEF) in the third week of August, under the command of Field Marshal Sir John French, while in general the overseas-based battalions arrived in France from September onwards. As it turned out, 1st KOSB did not deploy to France but returned to England where it remained until the beginning of 1915, when its parent 29th Division was sent to the Gallipoli front (see Chapter Seven).

At the outbreak of war Field Marshal Lord Kitchener was appointed Secretary for War, and at his first Cabinet meeting he astonished his colleagues by claiming that the war would last for a minimum of three years and would require over one million men to win it. On 8 August the call went out for the first 100,000 volunteers who would form the first of the New Armies. Kitchener's methods were as controversial as his prognosis was pessimistic. Instead of making immediate use of the part-time soldiers of the Territorial Force which had been raised for home defence in 1908 he decided to build on the existing regimental structure of the British Regular Army. No new formations would be raised but the existing infantry regiments would increase their numbers of battalions to meet the demand for men. These would be known as 'special service battalions' and the men who joined them would volunteer for the duration of the war. In that way, argued Kitchener, the volunteers could be assimilated quickly into the 'New' or 'Kitchener' armies and no new machinery would have to be assembled to deal with them. (For example, one officer in 6th KOSB received an immediate commission because he had been a member of the Malay States Volunteer Rifles while working in the Far East.) Although Kitchener placed little faith in the soldiers of the Territorial Force, dismissing them as a 'town clerk's army' full of sky-larkers, the existing battalions were allowed to volunteer for service overseas. Once introduced, on 13 August, the idea caught on and group pressure made it difficult for serving Territorials to refuse to serve overseas if the majority of the battalion volunteered. Both the KOSB first-line Territorial battalions served abroad with 52nd (Lowland) Division in Gallipoli, Palestine and on the Western Front, while the Regular battalions and the three service battalions served mainly on the Western Front (1st KOSB was also in Gallipoli) and 2nd KOSB shares the distinction with 2nd Gordons and 10th Black Watch of serving on the Italian front in the latter stages of the war. During the period 1914 to 1919 The King's Own Scottish Borderers expanded considerably to produce the following Regular, Territorial and Special Service battalions:

1st Battalion (Regular Army), Lucknow Brigade, 8th Indian Division, Lucknow, served with 87 Brigade, 29th Division

2nd Battalion (Regular Army), 13 Brigade, 5th Division, Dublin

3rd (Reserve) Battalion (Regular Army), Dumfries

1/4th (Border) Battalion (Territorial Force), Galashiels, served with 155 Brigade, 52nd (Lowland) Division

2/4th (Border) Battalion (Territorial Force), Galashiels

3/4th (Border) Battalion (Territorial Force), Galashiels

1/5th (Dumfries and Galloway) Battalion (Territorial Force), Dumfries, served with 155 Brigade, 52nd (Lowland) Division

2/5th (Dumfries and Galloway) Battalion (Territorial Force), Dumfries

3/5th (Dumfries and Galloway) Battalion (Territorial Force), Dumfries

6th (Service) Battalion (New Army), Berwick-on-Tweed, served with 28 Brigade, 9th (Scottish) Division

7th (Service) Battalion (New Army), Berwick-on-Tweed, served with 46 Brigade, 15th (Scottish) Division

8th (Service) Battalion (New Army), Berwick-on-Tweed, served with 46 Brigade, 15th (Scottish) Division

9th (Reserve) Battalion (New Army), Portland

10th (Service) Battalion (New Army), France

1914

Western Front: 2nd Battalion

Mobilisation of the 2nd battalion caused not a few problems as over 700 reservists had to be moved to Dublin from the depot at Berwick-on-Tweed, but despite the logistical difficulties the move took only four days. Embarkation began on 13 August, with the battalion travelling from Dublin to Le Havre on board the Bibby Line passenger ship SS *Gloucestershire*. The other battalions in 13 Brigade were 2nd Duke of Wellington's Regiment, 1st Queen's Own Royal West Kent Regiment and 2nd King's Own Yorkshire Light Infantry, all under the command of Brigadier-General G. J. Cuthbert. Once in France the 5th Division deployed north of the Mons-Conde Canal with 13 Brigade on the right, occupying the line St Vaast–Gommegnies. From there the division marched north into Belgium towards Mons, described by the regimental historian as 'a poor battlefield', but it was one that they had to defend by force of circumstance. The speed and aggression of the German attack led to the defeat of the French Fifth Army on the Sambre, a move that left the BEF isolated in their pre-arranged positions. Bearing down on them from the north were six divisions of General Alexander von Kluck's First Army. Sir John French agreed to hold the position for 24 hours and his men began digging in for the expected onslaught, using the features of the mining area, with its spoil heaps and buildings, to enhance their defences.

The expected German assault began on the morning of 23 August, and for the attacking enemy infantrymen it was a sobering experience. Trained to fire 15 rifle rounds a minute, the British regiments poured their fire into the advancing German lines with predictable results – the rate was so rapid and concentrated that the Germans believed they were facing machine-gun fire. When the élite 12th Brandenburg Grenadiers attacked 2nd KOSB in their position along the line near Lock 4 they presented irresistible targets 'in solid blocks, standing sharply against the skyline' (Brigade War Diary) and their casualties reflected the lack of tactical guile. Serving with the Brandenburgers as a company commander was the future novelist Walter Bloem and later, in his account of the fighting, he lamented the loss of his 'proud, beautiful battalion . . . shot down, smashed up – only a handful left . . . our first battle is a heavy, unheard-of heavy, defeat, and against the English [*sic*], the English we laughed at'. By the end of the day the attack had faltered as exhausted and frightened Germans attempted to regroup, but despite halting the assault the BEF was obliged to retire towards the River Marne and in the coming days its regiments were to receive increasingly high casualties. During the operation orders did not always get through to everyone and the bulk of C Company was obliged to surrender after being

The raising of the Earl of Leven's Edinburgh Regiment in the Abbey Close, Edinburgh, March 1689. Painting by Terence Cuneo.

A group of soldiers of the 25th during the regiment's tour of duty in Minorca, 1771. Painting attributed to Giuseppe Chiesa.

Operating in wintry conditions in Canada in 1866, when the 25th was involved in operations against Fenian infiltrators.

An old soldier talking to some of the younger generation in the guardroom of The King's Own Scottish Borderers, 1895. Painting by Frank Wood.

The North-West Frontier of India and Afghanistan was the scene of a succession of small wars in the later Victorian period. Gunner of 2nd KOSB, Chitral, 1895.

Riflemen and piper of 2nd KOSB during the Tirah campaign, 1897.

During the Boer War, part-time militiamen of 3rd KOSB were mobilised for active service.

Your King and Country Need You. Men of 4th KOSB prepare for service in 1914.

Borderers refilling gun belts in the trenches at Gallipoli.
Three battalions of KOSB took part in the campaign.

With the end of the war in sight, men of 2nd KOSB
relax at Arrewage, 1918.

Vehicles being served by men of 1st KOSB in Palestine in the 1930s, where the regiment formed part of the pre-war garrison.

During the Second World War, 7th KOSB was trained as an air landing battalion and saw action at the ill-fated Battle of Arnhem in September 1944.

Infantrymen of 4th KOSB in action during the fighting to take Flushing in 1944. Trained as mountain troops, they first saw action below sea level.

The Lady Mayor of Berwick inspects KOSB National Servicemen at their passing-out parade, the Barracks, Berwick-on-Tweed, May 1952.

During the Korean War, 1st KOSB served in the Commonwealth Division. In that faraway war, food parcels from home were always welcome.

A section of 1st KOSB on operations in the Radfan, the mountainous border region adjacent to Yemen, 1963.

The long war: patrol of 1st KOSB in Northern Ireland. The deployment, Operation Banner, lasted from 1969 to 2007.

Following the US-led invasion of Iraq in 2003, Basra became the responsibility of the British Army. Men of 1st KOSB patrol streets in Maysan Province.

surrounded and outnumbered. Particularly unfortunate was the officer commanding, Major A. E. Haig, who had been wounded in the shoulder and awoke from the anaesthetic to find himself in enemy hands.

The Great Retreat towards the River Marne, as it was known, took the German Army to the outskirts of Paris and the BEF suffered further casualties on 26 August, when II Corps turned to face the advancing Germans at Le Cateau some 30 miles from Mons. It was the British Army's biggest set-piece battle since Waterloo and their 55,000 soldiers faced German opposition which numbered 140,000. Lieutenant-General Sir Horace Smith-Dorrien's three divisions, supported by the Cavalry Division, were able to hold the line by dint of their superior fire-power but by evening they were outnumbered and only a German failure to press home their advantage allowed II Corps to resume their retreat. Even so, the casualties were heavy – 7,812 killed – and gave a stark indication of worse things to come. Exhausted by the battle and the summer heat, the BEF continued to pull back amidst rumours that the war was lost and that the French government had evacuated Paris for Bordeaux. It was a time of confusion, when the fog of war seemed very real indeed as the battle-weary infantrymen continued to sleep-walk through the French countryside – as Captain R. V. Dolbey, the medical officer of 2nd KOSB put it in his memoirs:

> In a dream we marched, unconscious of the towns we passed, the village we slept in; fatigued almost beyond endurance; dropping for sleep at the five minute halt that was the reward for four miles covered. All companies, dozing as they marched, fell forward drunkenly on each other at the halts; sleeping men lay, as they halted, in the roads and were kicked uncomplainingly into wakefulness again.

For the men of the KOSB ahead lay the next battle, on 13 September, when they reached the River Aisne where the long retreat ended and the Allies were able to counter-attack. This marked a new phase of the operations and signalled the end of a war of manoeuvre as both sides struggled to fill the gap between the Aisne and the Channel coast before it was exploited. This was known as the 'Race for the Sea' and it ended in stalemate with the only gap in the line being the wastes of the Flanders plain, an unprepossessing region peppered with names which soon become drearily familiar to the soldiers who fought over it – Ypres, Passchendaele, Messines, Langemarck, Vimy, Arras. The style of the fighting was also changing as the armies faced one another in the fields of Flanders. Trenches were dug, barbed wire obstacles were thrown up and field fortifications constructed; the German plan to encircle Paris had finally been blunted in the mud of Flanders and the first great set-piece battles were about to be fought.

1915

Western Front: 2nd, 6th, 7th, 8th Battalions

In 1915 the dilemma facing British and French planners was how to break the German trench line by attacking key points which would force the enemy to fall back on their lines of communication and in so doing return some fluidity to the fighting. Lines of advance had to be chosen and in January the Allies agreed to mount offensives against both sides of the German salient, which ran from Flanders to Verdun. These would be made at Aubers Ridge and Vimy Ridge to the north and in Champagne to the south, the intention being to squeeze the Germans and perhaps even converge to complete the encirclement of the salient. In this spring offensive the British and the French would attack in Flanders and Artois, the French alone in Champagne. For the British this involved them in battles at

Neuve Chapelle, Aubers Ridge and Festubert and later in the year at Loos. All failed to achieve the Allies' objectives and all produced large numbers of casualties. As a result the reality of trench warfare on the Western Front was brought home to the people of Scotland, especially at Loos in September where two of the six British assault divisions were the 9th and 15th Scottish Divisions of the New Army. Both divisions contained KOSB service battalions.

During the winter 2nd KOSB remained on the Ypres Salient where the regiment's war historian recorded the stark truth that any account of their existence would conclude that 'its revolting and dangerous monotony and physical discomfort do not invite analysis'. The battalion's first action of the year came in April, when the 5th Division was engaged in an assault on the German lines at a position known as Hill 60 on the Ypres–Commines railway line. Although it was taken, the Germans counter-attacked the KOSB positions and the battalion's casualties amounted to ten officers and 201 soldiers. This was followed by an attack on the Gravenstafel Ridge and St Julien which began the Second Battle of Ypres (22 April–25 May) when the Germans counter-attacked in an attempt to reduce the size of the salient. In both actions the battalion fought for the first time alongside Canadian troops, and it was also the first time that the men encountered gas, described in the regiment's war history as 'the phenomenon of a bluish, yellowish, whitish mist'. During the assault the Germans recaptured part of Hill 60 and only fierce Allied resistance kept them at bay, the *Official History* recording that 'shelled night and day from three sides, the conduct of the troops was magnificent'. In July 2nd KOSB moved out of the Ypres Salient for a new deployment further south on the Somme sector, a welcome move, not least because, according to the Brigade War Diary, the chalk downlands reminded the men of the landscape of Salisbury Plain.

By the summer the two Scottish New Army divisions had arrived in France bringing with them the 6th, 7th and 8th battalions of The King's Own Scottish Borderers, and their first taste of action came at Loos at the end of September. This was very much a Scottish battlefield. Of the 72 infantry battalions involved in the assault phase, half bore Scottish titles: not since Culloden in 1746 had so many Scots been involved in such a serious military undertaking. The battle was fought to support a French offensive in Artois and Champagne but the planning for the battle involved a great deal of compromise and in strategic terms its outcome was meaningless. Not for nothing was Loos known as 'the unwanted battle'. It began in the early hours of 25 September, when the British used gas for the first time and by mid-day there was optimism in the British ranks. The German line had been broken and there were reports of panic in Lens, where the headquarters of the German Sixth Army was making preparations to pull out. In some places there had been little opposition, a result of the gas and sustained bombardment, but already the Germans were rushing their reserves into the line and fighting hard to protect key points such as Hill 70 and the Hohenzollern Redoubt.

But those early successes had been bought at a terrible cost. As the Scots pushed their way through the smoke and gas with four battalions in line, each battalion split into three waves, the lead units of 26 Brigade suffered heavy casualties – 6th KOSB on the right lost 12 officers killed and seven wounded in the opening minutes and by the end of the day the numbers killed or missing amounted to 358, while 272 were wounded. As the battalion went into the attack they were encouraged by the piping of Pipe-Major Robert Mackenzie, a 60-year-old veteran who later died of his wounds, having been shot in both legs. In the same battle Piper Daniel Laidlaw, 7th KOSB, also piped his men into battle and as a result was awarded the regiment's first Victoria Cross of the war (see Appendix). The battalion reached the German lines as did 8th and 9th KOSB, but the battle was already slipping out of Haig's control. Now was the time to deploy the reserve divisions but due to poor planning they took time to arrive and the Germans were able to rush reinforcements into their threatened positions. By the time the fighting petered out in October the casualties told their own story: 7th KOSB lost 611 casualties out of a strength of 950 and 8th KOSB's casualties were 379 out of a similar strength.

1916

Western Front: 1st, 2nd, 6th, 7th, 8th Battalions

The year was dominated by the British offensive on the Somme, which began on 1 July and continued until the middle of November. The tactics produced by the new British commander-in-chief General Sir Douglas Haig were deceptively simple. He aimed to attack the German lines using the maximum force at his disposal, to break the defences and then to move forward to take possession of the area to the rear. To do this the British would attack with the Fourth Army numbering 19 divisions which would, in the words of the Tactical Notes produced for the battle 'push forward at a steady pace in successive lines, each line adding fresh impetus to the preceding line'. Following an enormous week-long bombardment involving the firing of a million shells along a 25-mile front the Germans would be in no condition to resist and the British infantry would simply brush the opposition aside as they took possession of the German lines. A creeping barrage would keep the surviving Germans cowering in their trenches, but alas it did not turn out like that. The opening day of the battle, 1 July 1916, produced the bloodiest day for the infantry regiments which took part in the initial attack. From the 11 divisions which began the assault 57,470 men became casualties – 21,392 killed or missing, 35,493 wounded and 585 taken prisoner.

Following the withdrawal of the Allied forces from Gallipoli (see Chapter Seven) 1st KOSB moved with 29th Division to Egypt prior to a new deployment on the Western Front where, according to the regimental war historian who served as the battalion's adjutant, 'those of us who went on leave envied the sheepskin coats, the enormous boots and British warms [overcoats] of those who were evidently far from strangers to France'. Their next destination was the Somme sector on the Hawthorn Ridge between Auchonvillers and Beaumont Hamel, where the men received intensive training in bayonet fighting and such novelties as cooperation with aircraft during operations. According to the regiment's war history, in the days before the battle the preparations were first-class and the men's morale was sky-high:

> The administrative branches and their technical advisers seemed to have made provision for everything. The rations were superb, and the canteens, incentives to extravagance, catered for every possible taste. Baths, dentists and dentures, gas-masks, steel helmets, waterproof sheets, concerts, troupes of entertainers were there for the health, safety, comfort and amusement of the men. And all this in spite of persistently squally weather.

Unfortunately all this splendid preparation was not matched by what happened on the first day. The bombardment went ahead as planned but it did not take long for the attacking soldiers to find that they were engaging well-defended positions and that the wire had not always been cut by the artillery fire. Accuracy was poor, far too many shells failed to explode and shrapnel proved useless in destroying the heavier barbed-wire defences. During the 29th Division's attack on Beaumont Hamel 1st KOSB was pinned down in open ground by accurate German machine-gun fire, leaving the regiment's war historian to note that 'the men on the spot had no alternative but to go on and be killed or wounded, or find cover'. In the fire-storm the battalion sustained 548 casualties and had to withdraw as best it could. Similar problems faced 2nd KOSB when it was committed to battle at High Wood in the third week of July. Failure of intelligence led to an under-estimation of the German strength and in the course of ten days the battalion lost 23 officers and 469 soldiers before being withdrawn to a quiet area south-east of Abbeville for rest and recuperation.

The Somme was also the battle in which many of the new service battalions were blooded and in most instances it proved to be a terrible baptism of fire. For 6th KOSB it was somewhat different as the battalion had fought at Loos, but that did not lessen the impact when it came. On the third day of the Somme 6th KOSB attacked German positions at Bernafray Wood to help consolidate the success achieved by the 30th Division in its capture of the village of nearby Montaubon. Half its bayonet strength became casualties during eight days of fierce fighting, 'a gruesome stroke [that] had to be taken passively'. Following a period of regrouping and retraining 6th KOSB returned to the Somme in October at a time of wet weather when the historian of the 9th (Scottish) Division recorded that 'all firmness had been soaked out of the ground, which became a sea of pewter-grey ooze'. Amongst other operations the battalion was involved in the attempts to take the Butte de Warlencourt, a position which would provide the British with better observation of the German lines. The site of a prehistoric burial ground, it dominated the surrounding countryside but as 6th KOSB found, its mass of tunnels and trench systems made it a formidable obstacle and it was not captured until the following year.

Even before the battle began the persistently high attrition rate on the Western Front had forced the amalgamation of the 7th and 8th battalions to form a composite 7/8th battalion and it was in action on the Ancre sector in mid-August when 15th (Scottish) Division attacked Martinpuich as part of the Battle of Flers. Nine officers and 280 soldiers became casualties following the assault, which captured five enemy trenches and a number of German machine guns during the first week of September. Described by the divisional historian as a well conceived and executed operation 'artillery, engineers and infantry worked together in a manner little short of miraculous; losses were not excessive, and a serious blow had been dealt to the enemy'. During the attack 7/8th KOSB was brigaded with 10th Cameronians, 10/11th Highland Light Infantry and 12th Highland Light Infantry. Flers marked the final period of fighting on the Somme for 15th (Scottish) Division and the rest of the year saw 7/8th KOSB engaged on what the regiment's war historian described as 'the serious but unpleasant duty of garrisoning the British front'.

1917

Western Front: 1st, 2nd, 6th, 7/8th Battalions

Italy: 2nd Battalion

The stalemate on the Somme forced the enemy to reappraise their options. Rightly fearing the renewal of a bigger Allied offensive in the same sector in the new year the German high command decided to shorten the line between Arras and the Aisne by constructing new and heavily fortified defences which would be their new 'final' position behind the Somme battlefield. Known to the Germans as the Siegfried Stellung and to the Allies as the Hindenburg Line, this formidable construction shortened the front by some 30 miles and created an obstacle which would not be taken until the end of the war. The withdrawal began on 16 March and as the Germans retired they laid waste to the countryside, leaving a devastated landscape in which the cautiously pursuing Allies had to build new trench systems. To meet this new challenge the Allies planned a new spring assault on the shoulders of the Somme Salient, with the French attacking in the south at Chemin des Dames while the British and Canadians would mount a supporting offensive at Arras and Vimy Ridge. Prior to the British attack, which began on 9 April, there was a huge and violent bombardment with 2,879 guns firing 2,687,000 shells over a five-day period.

In the opening phase of the battle both 6th and 7/8th KOSB achieved their objectives near the village of Monchy while 2nd KOSB supported the Canadians during their successful assault on Vimy Ridge north-east of Arras, where the Allies enjoyed the heady sensation of looking down onto the Douai plain and watching the Germans in full retreat. Casualties on the first day were one-third of those suffered in the comparable period on the Somme and large numbers of German prisoners had been taken. From the point of view of objectives being reached and casualties kept down, the first day of fighting at Arras deserves to be called a 'triumph'. Thereafter matters did not run so smoothly and the impetus was lost. Bad weather was one reason – the snow and rain did not make life easy for the men on the ground and delayed the transport – but it proved impossible to sustain the attack with exhausted troops. Any opportunity for an early breakthrough was lost when the Germans pushed reinforcements into the line and their arrival quickly nullified the earlier tactical advantage. By the time that the fighting ended at the beginning of May any hope of defeating the Germans at Arras had disappeared and the losses had multiplied. In a final attack on German positions on Greenland Hill 6th KOSB lost ten officers and 51 soldiers killed while a further 152 were wounded and 200 were listed as missing. 'The Borderers did their job, which was what one would have expected of them,' wrote the divisional historian, 'but few, very few out of those two companies who did their wheel and reached their objective ever got back. I think they were the only men who did reach their objective on the whole British front.'

During the Third Battle of Ypres, or Passchendaele, which followed in July 1917 and which was fought in an attempt to breach the Ypres Salient and reach the Belgian Channel ports 1st KOSB took part in the fighting at Langemarck, 6th KOSB attacked north of Gheluvelt, being forced to advance through the heavy mud which was a feature of the fighting, while 7/8th KOSB took part in the Battle of Pilckem. Like Arras, Third Ypres ended in failure with 250,000 casualties, many of whom were drowned in the muddy lagoons which littered the battlefield.

At the end of the year 2nd KOSB deployed to Italy with 5th Division as part of the British and French reinforcements sent to the country in the wake of the heavy defeat by Austro-German forces at Caporetto and the subsequent collapse of the home government. By the time the division arrived the situation on the Piave front had stabilised sufficiently to allow them to return to France in the following spring. The battalion War Diary recorded few items of interest and the regimental historian's verdict was that it was 'a pleasant interlude in the grim succession of operations on the Western Front'.

1918

Western Front: 1st, 2nd, 1/4th, 1/5th, 6th, 7/8th Battalions

The year opened with a major German offensive which began on 23 March and was to be the Germans' last attempt to defeat the Allies before the US Army brought much-needed reinforcements to the Western Front. (The US had entered the war in 1917.) The German strategy was brutally simple: the Kaiser's armies would drive a wedge between the two opposing armies, striking through the old Somme battlefield between Arras and La Fère before turning to destroy the British Third and Fifth Armies on the left of the Allied line. It almost worked too. By the beginning of April the Germans had advanced 20 miles along a 50-mile front, creating a huge bulge in the Allied line, and had pushed themselves to within five miles of Amiens. If this key city and railhead had fallen it would have been a disaster for the Allies. The French would have been forced back to defend Paris and the British would have been left with little option but to do the same to defend the Channel ports; the war would have hung in the balance. However, despite the obvious danger and the need to check it, the Germans had already shot their bolt by failing to concentrate the main thrust of their assault and dispersing the effort to take their targets.

As a result of the German attack the Western Front once again became a priority for the Allies and in April the 52nd (Lowland) Division was withdrawn from Palestine, landing in Marseilles before heading north to Abbeville to go into theatre reserve with General Sir Julian Byng's Third Army. It proved to be an unpleasant surprise for the men on board the transports: after the heat and dust of the Middle East they found themselves back in the familiar cold and wet weather of northern Europe and began a period of intensive training for the very different conditions of warfare on the Western Front. The division's machine-gun companies were reorganised into a machine-gun battalion and to meet the shortfall in manpower, as happened throughout the army, there was an internal reorganisation in the order of battle. Soon after arriving the division's strength was reduced by three battalions, which were sent to reinforce the badly depleted 34th Division. These were 1/5th KOSB, 1/8th Cameronians and 1/5th Argyll and Sutherland Highlanders.

The new alignment meant that all the KOSB battalions were fighting on the same front for the first time in the war. By that time conscription had been introduced (May 1916) and the complexion of the battalions had changed due to the constant reinforcement, but the character of the regiment was maintained by the presence of the 3rd Depot Battalion and the 9th (Service) Battalion through which most of the drafts passed. The former spent most of its war at the Marine Gardens in Portobello while the latter was based variously at Bordon in Hampshire, Dorchester, Stobs near Hawick, Catterick Bridge and finally at Kinghorn in Fife. During the last year of the war the KOSB battalions were involved in all the phases of the fighting, from repelling the German spring assault to the final offensive of 'the last hundred days' which brought the war to a successful conclusion in November 1918. Amongst the many operations in which they took part several stand out: the successful formation of a bridgehead at Marcoing on the Scheldt Canal by 1st KOSB during the Battle of Cambrai, when the British first used tanks in great numbers in November 1917; the defence of Kemmel Hill, south-west of Ypres, by 6th KOSB in April 1918 when the battalion lost 21 officers and 413 soldiers leading to the loss of two rifle companies; the involvement of 7/8th KOSB in support of the French Army at Buzancy during the Second Battle of the Marne in July 1918 where the losses were 32 killed, 214 wounded and 63 missing (mostly killed); the capture of Henin Hill on the Hindenburg Line on 26 August by 1/4th KOSB and the award of a Victoria Cross to Sergeant Louis McGuffie of 1/5th KOSB in the fighting at Wytschaete during the advance to the Ypres–Commines Canal; the final attack of the war undertaken by 2nd KOSB in the Selle Valley on 20 October 1918 'in the mist and the rain . . . to the tune of a mass of gun and TM [trench mortar] fire' when shortages of men had reduced the battalion to two rifle companies.

And then it was all over. At eleven o'clock on 11 November the armistice came into effect and the guns fell silent on the Western Front. For the soldiers the moment was met in several ways. Some celebrated, others were lost in quiet contemplation as they remembered fallen friends, but everywhere there was relief mingled with quiet satisfaction that the war had been won. An officer in the 15th (Scottish) Division overheard two men in 7/8th KOSB speaking about their feelings in subdued tones: said the first, 'I'd like fine to be in Blighty the nicht. It'll be a grand nicht this at hame; something daen' I'll bet.' 'Ay,' said another, 'an' there'll be a guid few tears, too.'

SEVEN

The First World War: Gallipoli and Palestine

The Western Front was not the only operational area for the British Army during the First World War. British soldiers also served in East and West Africa, Salonika (Thessaloniki), Mesopotamia (later Iraq), Palestine and Gallipoli. The latter three operations involved fighting against Ottoman forces and were part of a wider initiative to break the impasse on the Western Front by opening new fronts elsewhere. As it became clear that the war of movement was no longer an option on the Western Front by the end of 1914, so began the debate between the 'westerners' and the 'easterners'. The former argued that Germany could only be defeated convincingly in Europe while the 'easterners' believed that the stalemate could be broken by using the ships of the French navy and the Royal Navy to attack the Turkish positions in the Gallipoli peninsula at the mouth of the Dardanelles, the key to the Black Sea, and knock Turkey out of the war. Then ground forces would be landed to complete the capture of the peninsula and neutralise the Turkish garrison. Amongst the leading proponents of the concept was the

First Sea Lord Winston S. Churchill, who had made a name for himself as a war correspondent in Sudan and South Africa and was one of the rising stars of the ruling Liberal Party.

1915

Gallipoli: 1st, 1/4th, 1/5th Battalions

The naval plan to capture Gallipoli was put into operation on 19 February 1915 but it soon ran into trouble. Not only did the British and French battleships fail to make much impression on the Turkish defences but it had proved impossible to sweep minefields due to the accuracy of the Turkish field guns and the strength of the local currents. Several major warships were sunk or damaged and the decision was taken to land forces on the peninsula in an ill-conceived amphibious operation under the overall command of Lieutenant-General Sir Ian Hamilton, who had cemented his reputation as a fighting soldier during the Boer War. At the beginning of March it was agreed to earmark the experienced British 29th Division to support landings by light-infantry battalions of the Royal Naval Division. Kitchener then ordered the deployment of Australian and New Zealand troops training in Egypt while the French agreed to deploy the *Corps Expéditionnaire d'Orient*, a mixed force of French and North African soldiers.

The attack was planned to begin on 25 April, six weeks after the naval bombardment had spluttered to a halt. The main offensive was aimed at five beaches at Cape Helles, codenamed S, V, W, X and Y, and was directed by Major-General Aylmer Hunter-Weston, a Royal Engineer who had been born at Hunterston in Ayrshire in 1864. Known throughout the army as 'Hunter-Bunter' he was a hard and aggressive commander who frequently declared that he cared nothing about casualties provided that results were achieved; with his gruff red-faced ferocity he might have been a figure of fun had he not been in such an influential position. His 29th Division

('The Incomparable') was considered to be one of the best trained in the army and it contained the first of the KOSB battalions to take part in the Gallipoli theatre. But before they could give an account of themselves in the fighting they had to be taken ashore. That proved to be no easy task. In addition to a specially adapted Clyde-built collier the *River Clyde*, which was run ashore at V beach carrying 2,000 men, the main force was landed in ships' cutters pulled by a variety of tug boats. The intention was to secure the beaches and then to advance on the ridge between Krithia and Achi Baba, which provided the key to taking control of the peninsula. As was the case throughout the campaign things did not turn out that way.

In the initial stages the Turks seemed confused by the breadth and strength of the Allied attack but they soon regrouped and at V and W beaches the British forces took heavy casualties when they found themselves pinned down by heavy and accurate machine-gun fire. On the other hand the landings at X and Y beaches were unopposed, but there were serious communication failures between the two landing forces which meant that they were unable to exploit the situation even though they faced minimal Turkish opposition. Attacking Y beach was a combined force of 2,000 comprising 1st KOSB, the Plymouth Battalion Royal Marines and a company drawn from 2nd South Wales Borderers, but their progress was stymied by a muddle over who should take command. The KOSB commanding officer, Lieutenant-Colonel Archibald Koe, thought that he held command but no one had told him that the Marines' colonel was senior to him. The result was a fatal indecision which allowed the Turks to regroup and by nightfall the British forces were under heavy and sustained counter-attack. In a later letter to Hamilton which is contained in the Hamilton Archive at the Liddell Hart Centre for Military Archives, King's College, London, Captain Robert Whigham, a

KOSB company commander, reported that he could see the Turks approaching in the light of the full moon and described what he had observed:

> One could see line upon line of Turks advancing against our position. They fought with extraordinary bravery and as each line was swept away by our fire another one advanced against us and the survivors collected in some dead ground to our front and came on again. The attack worked up and down our whole front as if they were looking for some weak spot to break through our line. I saw one man, during one of these advances, continue to run towards us after his companions had stopped. He ran at full speed towards us, dodging about all over the place. He got up to within about fifty yards of the trench and then I saw him drop. Four times during the night they got right up to my trench before they were shot and one Turk engaged one of my men over the parapet with his bayonet and was then shot.

Forced to dig in quickly the Scots used their packs to reinforce their defences and later admitted that the 'trenches' never deserved the name. By the next day the Borderers' position was becoming untenable and both sides had taken large numbers of casualties; the British alone had lost 700 killed or wounded. With Hunter-Weston unable or unwilling to comprehend the seriousness of the position and the need for immediate reinforcement, a decision was taken to retire to the beaches for re-embarkation, a move which was covered by 1st KOSB. Koe was killed during the fighting and his body was never recovered. The failure to act decisively at Y beach was compounded by Hamilton's unwillingness to intervene in Hunter-Weston's direction of the battle and according to the

Official History those blunders typified the operation with its confusion over command, the lack of initiative after landing and the absence of any support from the staff:

> Cleverly conceived, happily opened, hesitatingly concluded, miserably ended – such is the story of the landing at Y beach. In deciding to throw a force ashore at that point Sir Ian Hamilton would seem to have hit upon the key of the whole situation. Favoured by an unopposed landing, and by the absence of any Turks in the neighbourhood for many hours, it is as certain as anything can be in war that a bold advance from Y beach on the morning of 25th April must have freed the southern beaches that morning, and ensured a decisive victory for the 29th Division. But apart from its original conception no other part of the operation was free from calamitous mistakes, and Fortune seldom smiles on a force that neglects its own opportunities.

Within the KOSB the feeling was that they had been left to their own devices and the high command had failed to respond to urgent signals. The battalion and the South Wales Borderers were in action again in the fighting for the village of Krithia on 28 April, but even at that early stage stalemate had come to the battlefield. Following a bombardment which lasted only 30 minutes due to shortages in ammunition the intention was to break out of the beach-head through a position known as Gully Ravine towards the heights of Achi Baba and push back the Turks from the tip of the peninsula but as the new commanding officer, Brevet Lieutenant-Colonel A. J. Welch, remembered later (for the regimental war history), through no fault of the KOSB the attack was badly planned – orders were issued late – and this led to blunders in its execution:

> By the time the remnants of the battalion reached a point about 900 yards from Krithia and could actually see into the village – not then occupied by hostile troops and therefore to be had for the asking – the fighting troops lacked officers; in other words, leadership. They consisted of KOSB and almost every unit of the 87th and 88th IB [Infantry Brigades], and were one and all in fine fighting form. 'Thank God! Here's an officer. What can we do now, Sir?' said one corporal of another unit to me. He had not seen an officer since early morning, i.e., soon after 8 a.m. It was a strange situation that met our eyes. Apparently on the R[ight] the French were in full retreat. Away on the L[eft] on the Aegean side of Gully Ravine the Borderers were being held up by MG [machine-gun] fire. And there we were in the L centre, opposite the open door, but without reserves other than our own depleted D [company]. Then busy digging in to protect our R flank. If *one fresh brigade* [author's italics] had been available, far-reaching results might have been gained.

During the fighting Welch was wounded; he was the third KOSB commanding officer to become a casualty and as a result Captain Whigham was placed in command. Despite the setbacks and the lack of progress the fighting continued and the incident became known as the Second Battle of Krithia. Once again the bombardment was modest and was further hampered by the lack of accurate intelligence about the Turkish positions. Once again, too, the enemy put up spirited resistance and by 9 May Hamilton was forced to admit that his plan had failed to work. The Allies were confined to their beach-head while the Turks held on to the higher ground and could not be dislodged, largely due to the doggedness of the opposition and the lack of adequate field artillery. At the same time the Turks failed to drive their enemy back into the sea and the fighting degenerated into as bitter a struggle as anything seen on the Western Front. By the end of the month, less than a week after they had landed, the British had lost some 400 officers and 8,500 other ranks (killed, wounded or missing), around one-third of the attacking force. In return, all they had gained was 600 yards of enemy territory.

At the beginning of May the first reinforcements had started arriving in the shape of the 29th Indian Infantry Brigade and the 42nd (East Lancashire) Division, but it was clear that Hamilton needed a much bigger force to dislodge the Turkish defenders. As a result, on 10 May Kitchener sanctioned the dispatch of the 52nd (Lowland) Division which had been training in the Stirling area. The division began deploying immediately, with the battalions being sent south by train to Liverpool and Devonport for passage to Gallipoli through Mudros or Alexandria. Sailing on board the Cunard liner *Mauritania* 1/5th KOSB arrived at Mudros in the late afternoon of 6 June and was taken ashore by naval motor torpedo boats landing at V beach ' . . . right in the midst of a fearsome, deafening pandemonium, which was to last, with little interruption, during the seven months of our stay on the Peninsula' (War Diary). Six days later, while supporting the 1st Naval Brigade, the battalion lost its first casualty of the campaign – Private Robert Teesdale of Dalbeattie who was killed by shellfire.

By the time that the replacements of 52nd (Lowland) Division arrived 1st KOSB had been in action again, fighting in the Third Battle of Krithia, which was in effect a reprise of the attempts to take the Achi Baba positions. To the 29th Division fell the task of clearing the Turkish trenches on either side of Gully Ravine, but when the attack began in the morning of 4 June the assault formations soon found that the preceding bombardment had done nothing to silence the opposition's machine guns. As 2nd Lieutenant Richard Reeves noted in his diary (held by the Imperial War

Museum), when the battalion formed the second wave of the assault with 4th Worcesters the men were simply sitting ducks and the casualties were correspondingly high:

> At 12 punctually the intense bombardment ceased – it was an infernal noise – no words can describe the hideous din – the earth simply shook & parts of the support trench in which I was fell in from the reverberation. A and B Coys attacked and lost very heavily . . . C & D Companies followed 50 paces behind, and we had to get up a very high parapet in the face of a perfect hail of shrapnel and machine-gun and rifle fire – I ran on blindly shouting to my men – we lost heaps – men falling around me and with such terrible wounds . . .

Although 1st KOSB and 4th Worcesters succeeded in taking their objectives and took a large number of Turkish prisoners, the assault as a whole was a failure, the regimental war history describing the gains on the 4,000-yard front as 'paltry'. Once again, uncut wire was largely to blame for the British casualties, which amounted to 4,500 killed, wounded or missing, and hopes for a quick breakout were dashed. Even so, Hunter-Weston was determined to press on with the attempt to dislodge the Turks and this led to a fresh assault on 28 June in which 29th Division was given the task of attacking heavily wired enemy positions which had managed to resist all attempts to bombard them into submission. Also taking part in the attack was 156 Brigade, part of the recently arrived 52nd (Lowland) Division. On the flanks the attack achieved some success and as a result Hunter-Weston decided to concentrate on the centre to bring it into line and perhaps even provide the impetus to push back the Turks once and for all. To do that the 52nd (Lowland) Division was ordered into the line when the attack began on 12 July, with the French on the right with 155 Brigade. Following the usual preliminary barrage the attack opened at four-thirty in the morning with 1/4th KOSB in the lead, and in the initial stages of the assault the Scots managed to reach the first- and second-line Turkish trenches with relatively few losses. However, in the confusion of the fighting many men found themselves too far ahead of the attack and fell into enemy hands. Amongst them was a private in 1/5th KOSB whose story is related in Scott Elliot's war history:

> Men fell like corn below the scythe . . . I managed to get to the furthest point, that was the third Turkish trench or dummy trench. It was about one foot deep, and we had to set-to and fill sandbags. We were packed together and enfiladed from the left. Our fire rapidly diminished, till there was no one else left to fire. Then I was knocked out. When I came to, our little trench was occupied by a Turk to every two yards. Four or five of our men were laying across me, and I could not get up. I was bayoneted six times in the back whilst lying there. A Turk officer, at the point of his revolver, ordered the Turks to release me.

By nightfall the men of the Lowland Division had advanced some 400 yards and it was not until 14 July that the two KOSB battalions were relieved. Of the 900 or so men of 1/4th KOSB who had gone into the attack 12 officers had been killed and six were wounded, while 319 soldiers had been killed and 203 had been wounded. Amongst the casualties were the commanding officer and the adjutant. The losses were only marginally better in 1/5th KOSB: six officers killed and five wounded and 76 soldiers killed and 183 wounded. During the operations of 21 and 28 June and 12 July the total British losses were 7,700. As for the Turks, they simply consolidated their position and built fresh lines of trenches.

By then the fighting for position at Helles had reached stalemate and the conditions facing the Allied troops had sharply deteriorated. Not only was the fighting conducted at close-quarters, with some trenches being almost in touching distance, but the physical hardships were worse than anything faced on the Western Front. Despite the best efforts at maintaining basic sanitation disease was rampant, especially dysentery and enteric fever, which were spread by the absence of proper latrines and washing facilities and by the ever-present swarms of black buzzing flies. One medical officer said it was impossible to eat in their presence as they quickly swarmed onto any spoonful between plate and mouth. Like everyone else on the peninsula the Scots had to live with the plague, although as George Waugh, 1/4th King's Own Scottish Borderers, recalled in an interview now held by the Imperial War Museum, some even managed to turn the flies into a joke: 'There was raisins and currants he [the cook] shoved in the rice. But this day when they were dishing it out there was no currants in it. One of the men beside me said, "Nae currants, Davie?" The cook replied, "Nae currants today. Oh, half a minute." He lifted the Dixie lid and all the flies went in!' In the heat of high summer the swarms were especially bad, and even the advent of colder autumn weather brought little respite as the sun gave way to long days of freezing rain.

Despite the arrival of fresh reinforcements – in all, Hamilton was given five new divisions – the deadlock could not be broken and the men on the peninsula were becoming increasingly weakened. An ambitious amphibious landing at Suvla Bay failed in August because the Turks were able to rush reinforcements into the area to prevent the creation of a bridgehead. The news of the failure was met with dismay in London and compounded the idea that 1915 had been a year of military disasters, with no change in the position on the Western Front and stalemate in Gallipoli. Disease and searing heat added to the difficulties facing the men on the peninsula and in August alone over 40,000 soldiers had to be evacuated, the majority suffering from dysentery.

The strategic situation worsened the following month when Bulgaria began moves to enter the war on the opposite side and an alarmed Serbia made urgent requests for British and French reinforcements. In October the inevitable happened: Hamilton was sacked, rightly so as his leadership had become increasingly feeble and sterile, and he was replaced by General Sir Charles Monro, a veteran of the fighting on the Western Front who was also a disciple of the westerners. Having taken stock of the situation he recommended evacuation, although this was not accepted until the beginning of November, when Kitchener himself visited the battle-front and found himself agreeing that the difficulties were insuperable. A heavy and unexpected winter storm also helped to decide the issue – over 280 British soldiers died of exposure, including a number from the Worcesters who were found frozen to death on the fire-steps of their trenches. In a brilliant operation, which was all the more inspired after the fiascos which preceded it, the British finally withdrew their forces at the beginning of 1916, remarkably without losing any casualties. Captain Stair Gillon served in 1st KOSB and his account of the withdrawal reveals the mixture of relief and regret which accompanied the operation as the men made their way to the beaches and safety:

> Patience, and one trod a gangway. A heave and a hoist on to the deck and a pleasant invitation to come downstairs. One had seen one's last of Gallipoli. One last excitement – a terrific explosion followed by a thunderous shock above our heads. It was the big dump being destroyed. Soon after, nautical noises and rattling chains were followed by the soothing shudder of the screw and the unmistakable roll

> of the sea. Perhaps the Turk thought we couldn't get away that night and was reserving his shells for a more likely occasion.

The great adventure to win the war by other means was finally over. As happened on other fronts the exact British death toll was difficult to compute, but most estimates agree that 36,000 deaths from combat and disease is not an unreasonable tally. (The official British statistics show 117,549 casualties – 28,200 killed, 78,095 wounded, 11,254 missing. Total Allied casualties are put at 265,000.) The KOSB casualties are also difficult to quantify, the regimental historian contenting himself with the thought that on the theory of averages the figures might be divided amongst the three battalions which served in Gallipoli throughout the campaign.

The failure of the Gallipoli campaign has provided history with one of its great conundrums, the conditional 'if only' being applied to most aspects of it. If only the tactics, the leadership, the reinforcements and the munitions had been better, if only the execution had matched the conception, then a sordid defeat could have been a glittering triumph. The original reasons for the deployment had much to recommend them, but an absence of clear thinking and the half-hearted conduct of the campaign must account for its failure and for the waste of so many lives and so much equipment. The *Official History* summarised the campaign in words that spoke only of disillusionment and despair:

> Few memories are sadder than the memory of lost opportunities, and few failures more poignant than those which, viewed in retrospect, were surely avoidable and ought to have been avoided. The story of the Dardanelles is a memory such as these.

The failure of the campaign left a lasting bitterness amongst those who had taken part, not least in Australia and New Zealand, where 25 April is still commemorated as Anzac Day. Unlike the Western Front, where optimism survived for a surprisingly long time, there were no good words to be said about Gallipoli and soldiers who served on both fronts admitted that the conditions on the peninsula were worse than anything they encountered in France and Flanders. However, it was not the end of the war for the survivors. The original three divisions – 29th, Anzac and Royal Naval – were sent to France in time to take part in the Battle of the Somme and were followed by the 11th (Northern) and 42nd (East Lancashire) Divisions after a brief period of rest in Egypt. The 13th (Western) Division was despatched to Salonika to assist the Serb war effort and the rest of the Mediterranean Field Force, including the 52nd (Lowland) Division, remained in Egypt to guard the Suez Canal and, later, to take part in new operations against Ottoman forces in Palestine and Syria. With them went the two KOSB Territorial battalions.

1916

Egypt: 1/4th, 1/5th Battalions

By the time the Gallipoli forces reached Egypt the immediate threat of an Ottoman attack on the Suez Canal had receded but the strategic situation was still parlous. In addition to the withdrawal from Gallipoli British prestige had also been dented by the surrender to the Turks of a British and Indian force at Kut-al-Amara as part of an operation to gain control of the oil-rich province of Mesopotamia. Following a sybaritic interlude on Lemnos 52nd (Lowland) Division moved to Egypt at the beginning of February and was allotted to VIII Corps which guarded the southern sector of the defensive line running from the Red Sea to

Kabrit on the Great Bitter Lake. Overall command of the British and Imperial forces in Egypt was in the hands of General Sir Archibald Murray, formerly Chief of the Imperial General Staff. For the most part this was a quiet period for both battalions and days were dominated by training and learning new techniques in an area described by the regimental war historian as a sojourn in the wilderness remembered for the 'starry nights strangely spent by kindly Scots'. There was also work to be done guarding the efforts to push the railway line and the fresh water pipe north towards El Arish in preparation for the anticipated attack on Ottoman forces in neighbouring Palestine. One memory above all others stands out from this period: learning to march on the desert sands below the heat of the burning sun (the description comes from Scott Elliot's history):

> The battalion marches with fours well spread out in width but not in length, so as to minimise dust and allow free passage of the air; all ranks are in short-sleeves, with the lower arms bare and the neck open; the drill jacket is strapped to the haversack, or inside the pack if the latter is being carried, but pack and haversack were never worn together while we were on trek, one or other being left behind to be brought up later. The company commander is probably walking at the head or by the side of his company. At the rear of each company are its pack animals, carrying its Lewis guns and a proportion of ammunition.

Learning to march on sand was likened to walking on shingle or deep snow, and the speed was never more than one and a quarter miles per hour as opposed to the usual three miles an hour on metalled roads. For some there was the satisfaction of knowing that they might have been following in the footsteps of Alexander the Great or Napoleon 'not to mention Abraham and Jacob,

Joseph and his brethren and the Holy Family' but most soldiers remembered only the tedium, the heat and the dust and the desire for limitless supplies of cold running water. The commanding officers of 1/4th and 1/5th KOSB were, respectively, Lieutenant-Colonel M. B. Saunders and Lieutenant-Colonel J. R. Simpson, formerly Highland Light Infantry.

1917

Palestine: 1/4th, 1/5th Battalions

Following the failures in Gallipoli and Mesopotamia, the next strategic aim for the British was the defeat of the Turkish forces in Palestine. In January the campaign opened with the fall of Rafa, which opened the route towards Gaza, the gateway to Palestine and home to a Turkish garrison. The first attack on the position took place at the end of March and it seemed to be succeeding before the cavalry was withdrawn in the mistaken belief that the infantry assault had failed. Murray ordered a second attack on 17 April, but it too faltered in the face of heavy Turkish fire. During the attack on a position called Ali el Muntar the two KOSB battalions lost a total of 500 men, including a number of experienced officers. All told, the British casualties were 6,000, while the Turks lost one third of that number. As a result of the failure Murray was sacked and replaced by General Sir Edmund Allenby who had already been dismissed after his failure at Arras earlier in the year. Before taking up his new command in June he had been warned by the new prime minister, David Lloyd George, that he had to take Jerusalem by Christmas as a gift to the British nation and that he should demand what he needed to make sure the enterprise succeeded. In fact, there was already a pressing need to attack the Turkish forces, who were being reinforced in Aleppo, in present-day Syria, for an offensive which would be known as *Yildirim*, or lightning thrust, to retake Baghdad. If Allenby could engage the enemy through Palestine it would force the Turks to divide their forces and pass the initiative back to the British. There were other imperatives. With no sign of a breakthrough on the Western Front Lloyd George hoped that the defeat of Turkey would be a major blow to the Germans and perhaps hasten the end of the war without further costly offensives in France and Flanders. From a strategic standpoint the British and the French had already drawn up secret plans to carve up areas of influence in the Middle East following the expected collapse of the Ottoman Empire.

From the outset Allenby recognised that he needed overwhelming superiority over the Turks if he were to avoid the setbacks of Gallipoli and Kut, but getting the reinforcements in the second half of 1917 was another matter. The priority continued to be the Western Front, and with the Battle of Passchendaele eating up resources it took time and much subtle diplomacy to build up his forces. Eventually these consisted of a Desert Mounted Corps made up of Anzac and British cavalry, yeomanry regiments and two infantry corps, one of which (XXI Corps) contained the 52nd (Lowland) Division, which now included the Lowland Mounted Brigade (1/1st Ayrshire Yeomanry). Their objective was to break into Palestine through Gaza and Beersheba and destroy the defending Turkish Eighth Army under the command of Friedrich Freiherr Kress von Kressenstein. Allenby had also requested and received modern warplanes and the use of DH4 bombers to carry out air strikes was central to his plans, which were based on hitting the enemy hard and then deploying mobile forces, including cavalry, to take them by surprise.

Allenby's battle-plan was innovatory yet simple, and was based largely on his reading of the terrain over which the fighting would take place. On 27 October, following a huge bombardment from artillery and from British and French warships lying off the coast, Gaza was attacked for the third time while XXI Corps prepared for the break-in battle which would follow. This confirmed to the Turks and their German advisers that Allenby was using the tactics of the Western Front, and that assault would come from the waiting ground troops. However, at the same time and in great secrecy Allenby shifted the emphasis of his attack towards Beersheba, which was quickly surrounded by a brilliant flanking attack to the east. At the end of a battle which lasted all day on 31 October the issue was decided with a full-scale cavalry charge into the town carried out by the 4th Australian Light Horse. Not only had the objective been taken, but Beersheba's vital water supplies had been captured and the Turks were powerless to bring up reserves. The fall of Beersheba allowed Allenby's forces to put greater pressure on the Turkish positions at Gaza, and its defences were successfully stormed on 7 November, leaving the defenders no option but to retreat north up the coastal littoral towards Askalon and Jaffa. During this mobile phase of the battle the Turks fought with great determination, but they were demoralised by the weight of the attack and by the use of British warplanes to strafe their fleeing columns. Although there were concerns in London that Allenby might repeat Townshend's mistakes during the Kut operations by overstretching his lines of communication, the strength and aggression of the British attack brought his army to the western approaches to Jerusalem by the end of November. This took 1/5th KOSB into hills which reminded the men of the Border hills of home, but the scholar in Scott Elliot saw an older landscape which he recorded in a footnote in his battalion's war history:

> The whole of this countryside is celebrated in history. Surely the spirits of Joshua, who drove the armies of the Five kings like a flock of sheep over this very battlefield; of Judas Maccabaeus, who had hunted an enormous host of Syrians down the Bethhorons; or many a brutal but

> singularly successful Roman General; of Saladin the Sultan and of Richard Coeur de Lion, who had struggled with each other – surely they were watching in admiration and astonishment this deadly struggle, waged with weapons more destructive than any they had ever dreamed of.

To great acclaim the holy city fell on 8 December after a determined attack by the 53rd and 60th Divisions forced the remaining defenders to evacuate their positions. Three days later, in a carefully stage-managed operation, Allenby and his staff entered the city to take possession of it and to secure the holy places. To avoid hurting Islamic feelings the Mosque of Omar was put under the protection of Indian Muslim troops, while the guards lining the streets came from the four home countries, Australia and New Zealand and France and Italy.

It was not the end of the war in Palestine, but it was the beginning of the end. Allenby's next objectives were to move into Judea and to regroup to prevent Turkish counter-attacks before moving on to his next objectives, Beirut, Damascus and Aleppo. However, to accomplish that he would need additional troops to reinforce his own men and to protect the lines of communication as he pushed north; at the very least, he told the War Office, he would need an additional 16 divisions, including one of cavalry. In the short term his forces invested Jaffa, which fell after the 52nd (Lowland) Division seized the banks of the River Auja in an operation which demanded surprise and resulted in what the divisional historian described as 'the most furious hand-to-hand encounters of the campaign'. Finding their way blocked by Turks, who fought to the last round, the men of 1/5th KOSB 'drove them towards the river, and as the Turks would not surrender, into it'. As a result 30 Turks drowned, 15 were bayoneted and 20 were taken prisoner.

This proved to be the last action undertaken by the KOSB

battalions in Palestine. Before the question of reinforcing Allenby could be addressed by the War Office the Allies were faced by a crisis on the Western Front in March 1918, when a German offensive pushed back the British line and almost led to its collapse (see Chapter Six). During the fighting the British sustained 163,000 casualties; the need for rapid reinforcement had come at the very moment when Allenby wanted to continue the push towards Aleppo and he was forced to order two infantry divisions, nine yeomanry regiments and one divisional artillery unit to move to France. One of the divisions was 52nd (Lowland), which embarked for Marseilles at Alexandria aboard seven troopships escorted by six Japanese destroyers. Later they were replaced by the two Indian divisions from Mesopotamia and the decisive battle took place at Megiddo on 19–21 September. After a fierce artillery assault infantry and cavalry pushed through the Turkish lines, harrying the enemy as they tried to escape and pushing north into the Plain of Esdraelon. The way to Damascus was now open and with Turkish authority collapsing the end of the campaign was in sight. On 31 October an armistice was signed. It was one of the cheapest and most successful campaigns of the war: in the space of five weeks Allenby's forces had advanced 300 miles and captured 75,000 Turkish soldiers and 360 guns for the loss of 6,000 casualties.

All told, some 50,000 Scots served in the four operations which were devised by the 'easterners' to produce an alternative way of defeating Germany. The expected breakthrough never happened but Allenby's operations defeated the Turks and helped to settle, for good or ill, the post-war shape of the Middle East by which the former Ottoman provinces were divided up between Britain and France and a Jewish homeland was settled in Palestine. By far the bloodiest and most wasteful campaign had been Gallipoli, which was a story of avoidable failure and wasted opportunities, and the memory of the fighting on that barren peninsula lingered longest in the minds of the soldiers who fought there. In an open letter to the *Scotsman*, Dr William Ewing, a chaplain with the 52nd (Lowland) Division, said simply that 'the sights I have seen will never be erased from memory' and that the spectacle of so many wounded and mutilated men provided 'some conception of what this strife costs'. All told, the fighting against the Ottoman forces had been a long and unforgiving experience. The historian of the 52nd (Lowland Division) recorded that in three years of service in the Mediterranean theatre, the Scots had 'fought in all kinds of country, and in all weathers and temperatures, ranging from the hard frost and blizzards of the moorlands of Gallipoli, and the bitterly cold hail storms of the Judaean Highlands, to the mid-day heat of summer in the desert'.

EIGHT

The Second World War: France and North-West Europe

With the conclusion of hostilities in 1918 it was inevitable that there would be cutbacks in defence expenditure and that the huge wartime army would be gradually dismantled. In the KOSB the three New Army battalions were disbanded, but not before the 6th battalion had the honour of being one of the formations selected to march into Germany as part of the army of occupation. A few days before the armistice came into being 7/8th KOSB had the novel experience of passing close to the village of Fontenoy, where the Edinburgh Regiment had fought in 1745, and the 2nd battalion was also touched by past history. It ended the war in the vicinity of Namur, as did another regiment, 1st Bedfordshires (formerly 16th Foot), which had also taken part in the famous siege all those years ago. Both battalions took the opportunity to commemorate the incident by trooping their colours in the town, a historic moment which the battalion War Diary scarcely mentioned other than to record: 'Colour Party returned with Regimental Colours'.

For the two Territorial battalions the end of the war brought to a close their soldiers' wartime service. By chance 4th KOSB found itself east of Sirault and only six miles away from Lock 4 on the Ypres–Commines Canal where the 2nd battalion's high rate of fire had killed so many Brandenburgers in the opening weeks of the war. The 5th battalion also marched into Germany, but not before passing its first peacetime days close to the old battlefield at Oudenarde.

For the two Regular battalions there was a return to peacetime soldiering. The 1st battalion spent a short spell with the army of occupation in Germany before returning to its pre-war posting in India, where in many respects life had not changed all that much since the late Victorian period. Life was different for the 2nd battalion. It was one of many infantry battalions sent to Ireland to support the police in the civil war which had broken out there (known variously as the Anglo-Irish War or War of Independence) and which preceded independence in 1921. It was a squalid little conflict, with murders and revenge killings carried out by both sides, the Irish Republican Army (IRA) as well as the security forces. 'The whole country runs with blood,' said a leader in the *Irish Times* on 20 April 1921. 'Unless it is stopped and stopped soon every prospect of political settlement and material prosperity will perish and our children will inherit a wilderness.' A political solution was imposed on the country and its warring factions at the end of 1921, but the decision to keep the six Ulster counties separate from the Irish Free State was to cause equally vexing problems later in the century.

In 1922 1st KOSB moved to Egypt, where it was quickly rushed to the Dardanelles to form a buffer force between rival Greek and Turkish forces at Chanak (present-day Çanakkale) on the Asiatic side. The crisis subsided in 1923, allowing the battalion to return to Edinburgh for the first time in 141 years. This was followed by tours of duty at Aldershot, Fort George and Catterick and a deployment in Malta in 1935. For the rest of the decade the battalion was in Palestine as part of the British garrison attempting to keep the peace between the Arabs and the increasing numbers of Zionist settlers which flocked into the Holy Land as a result of the wartime decision to create a Jewish homeland. As tensions mounted between the rival populations in the 1930s, the British produced two partition plans aimed at providing a workable settlement for the Jewish and Arab populations. Neither succeeded in achieving their objective and that failure left the British Army holding the line as fighting broke out between the two communities. For 1st KOSB, and for every other battalion which served in Palestine, it was a difficult period, with the men having to try to keep the peace, usually without being able to identify the terrorists who were attacking them. By then 2nd KOSB had left its equally demanding posting in Ireland and moved to the more pleasing environment of Hong Kong where, the regimental historian records 'the sociability of the Officers' Mess became a feature of the Colony's British community'. In 1930 the battalion returned to India, where it was based throughout the decade at Poona, Lucknow and Calcutta. Later (see Chapter Nine) it would be based on the North-West Frontier, where it took part in internal security operations against the mettlesome Pathan tribesmen.

By then it was obvious to many that once more the world was heading towards war. In Germany the Nazis had come to power under Adolf Hitler and their presumptuous territorial claims were soon trying the patience of the rest of Europe. In 1938 Prime Minister Neville Chamberlain seemed to have bought 'peace in our time' following his negotiations with Hitler in Munich, which gave the Germans a free hand in the Sudetenland and subsequently in Bohemia and Moravia. However, it proved to be the calm before the storm. Having signed a peace pact with the Soviet Union, Hitler then felt free to invade Poland at the beginning of September 1939. Chamberlain, who would be replaced as prime minister by Winston Churchill the following year, had no option but to declare war – Britain and Poland were bound by treaty – but the country's armed forces were hardly in a fit condition to fight a modern war. The British Army could only put together four divisions as an expeditionary force for Europe, six infantry and one armoured division in the Middle East, a field division and a brigade in India, two brigades in Malaya [now Malaysia] and a modest scattering of imperial garrisons elsewhere. Years of neglect and tolerance of old-fashioned equipment meant that the army was ill-prepared to meet the modern German forces in battle and British industry was not geared up to make good those deficiencies. Once again in the nation's history, it seemed that Britain was going to war with the equipment and mentality of previous conflicts. Events in Poland quickly showed that Germany was a ruthless and powerful enemy whose Blitzkrieg tactics allowed it to brush aside lesser opposition: using armour and air power the Germans swept into the country, which fell within 18 days of the invasion, allowing Hitler to turn his attention to defeating France.

1940: BLITZKRIEG AND DUNKIRK

1st, 4th and 5th battalions

To bolster the French army in the opening months of the war the British government deployed a British Expeditionary Force (BEF) which included Regular and Territorial battalions of The King's Own Scottish Borderers. As part of 9 Brigade in 3rd Infantry Division the 1st battalion crossed over to France in September under the command of Lieutenant-Colonel E. E. Broadway and joined the British forces in positions near Lille on the border with Belgium. It was in good company: the divisional commander, Major-General Bernard Law Montgomery (later Field Marshal Viscount Montgomery of Alamein) referred to the formation as the 'International Brigade' – 9 Brigade was composed of English, Irish and Scottish battalions – 2nd Lincolnshire Regiment, 2nd Royal Ulster Rifles and 1st KOSB. This was a frustrating period which came to be known as the 'phoney war', when both sides considered their options in western Europe following the rapid collapse of Poland. To defeat the western Allies the German plan called for the invasion of the Netherlands, Belgium and Luxembourg, using two army groups to smash through the southern Netherlands and central Belgium while a diversionary attack was made through the Ardennes. The ultimate goal was control of the Channel ports as a prelude to invading Britain. However, Hitler prevaricated, the plans were subjected to constant change and there were delays in correcting the balance of ground forces. At the same time, the French dithered and ordered an unnecessary move into the Saarland which did nothing to alter the strategic balance in the Allies' favour and introduced a sense of demoralisation and defeatism. As for the British, they had deployed nine infantry divisions in France but there were deficiencies in armoured and artillery support as well as in air cover. Compared to the Luftwaffe's 4,200 war planes the Allies possessed only 2,000, half of which were fighter aircraft. In qualitative terms the German machines were also superior and their air crews enjoyed better training and superior tactics.

War came to the BEF with a vengeance on 10 May 1940 when the Germans subjected France to the frightening tactics of Blitzkrieg, using armour and air power to back a rapid ground assault into Belgium and the Netherlands. The 1st battalion moved immediately into Belgium to take up positions along the River Dyle to the east of Brussels, but this was only a prelude to a steady withdrawal as the BEF fell back under the weight of the German assault. On the third day 1st KOSB was deployed on the River Escaut, close to the old battlefield at Oudenarde, and with each passing day the British formations found themselves on the back foot as they started pulling back towards the Channel ports. Names from the earlier global war were passed – Ypres, Tournai, Armentières, Bailleul, Lille, Abbeville – and eventually the beaches at Dunkirk became the final destination. Once there, the only alternative to a last stand was evacuation, and as resistance would have resulted in the destruction of the British Army the BEF was able to pull out using an incredible mixture of naval craft and the famous 'little ships' which produced the 'miracle of Dunkirk'. All told, 338,226 soldiers made good their escape, thanks mainly to solid discipline (the retreat never became a rout), the gallant defensive battle fought by French forces at Lille and indecision on the part of the German high command. A Borderer, Henry Bridges, captured the mood in his poem 'Dunkirk', paying tribute to the armada of little ships taking part in the operation. It was published in the regiment's war history:

> The sands were black with soldiers, and the skies were
> black with planes,
> But the Monarchs and the Skylarks and the little Saucy
> Janes
> Undaunted by their danger set their course through
> the attack
> And from that hell of bloody death they brought our
> soldiers back.
> And women wept in England then with happiness
> and joy,
> And many an English mother ran to welcome home
> her boy.

Once back in Britain 1st KOSB reassembled at Shepton Mallet near Wells in Somerset. Its new commanding officer was Lieutenant-

Colonel D. C. Bullen-Smith, a future commander of the 51st (Highland) Division.

The need for recruits met a ready response in the regimental area, especially in the two Territorial Army battalions. In Hawick the 4th battalion had 300 soldiers on its strength and even small villages like Newcastleton had 80 men on parade. Other detachments were based at Ayton, Chirnside, St Boswells, Coldstream, Duns, Greenlaw and Lilliesleaf. The 5th battalion was no less industrious. Based in Dumfries, it had detachments across the western Borders and was commanded by Lieutenant-Colonel G. G. Walker. So desperate was the need for recruits that both the existing Territorial battalions spawned duplicate battalions – 6th KOSB and 7th KOSB. The regimental records show that the appeal was spread across the Borders: the 6th battalion, brother to the 4th battalion, was drawn from Jedburgh, Kelso, St Boswells and Newcastleton (all A Company); Hawick (B Company); Galashiels (C Company); Ayton, Duns and Chirnside (all D Company) while Headquarters Company numbered men from Hawick and Selkirk. An 8th battalion was also formed for home defence duties, its primary role being to guard the Portpatrick radio station and the seaplane base at Wig Bay. Later it served on Orkney; a 9th battalion was formed but did not see service. At the outbreak of war the 4th and 5th battalions provided home defence duties and to the latter fell the honour of being the first Borderers to fire shots in anger. Some idea of the urgency of those days can be seen in a memoir supplied by James MacQuarrie for the BBC's 'People's War' stories:

> Three days before war broke out, I was working at Lairdmannoch Loch with Jim Glover. We were painting and repairing the small rowing boats used for fishing on the loch. At dinnertime we went down to the big house to

> eat our piece. While we were there we heard over the radio that all territorials were to report immediately to their drill hall. Out of the five of us who had joined two years earlier I was the only one left.
>
> I reported to the drill hall at Gatehouse that afternoon along with the rest of the lads. We were told that a bus would pick us up on Sunday at 11 a.m. to take us to Dumfries. I played football for the Saints on Saturday as usual and on the Sunday went to Gatehouse, where the bus picked us up as promised and off we went.
>
> As the bus went by the Star Inn at Twynholm, Miss Carter, Mrs Lamont and my grandfather were standing outside wiping their eyes with their hankies. I wondered what was wrong with them.

After being billeted at Rosefield Mills in Dumfries the 5th battalion moved north to the Edinburgh area, where it was engaged in home defence duties, protecting Turnhouse Air Force base and the Forth Railway Bridge. On 16 October, while providing a guard for the Forth Railway Bridge, soldiers from the battalion opened fire on German bombers which were attempting to bomb British warships anchored off South Queensferry.

During the winter of 1939–40 both the 4th and 5th battalions formed part of 52nd (Lowland) Division, which was sent to France early in June 1940 as last-minute reinforcements, disembarking at St Malo on 13 June. It proved to be too little and too late. Although the Scots got as far as the Le Mans area the Germans had arrived in strength and the division was forced to retreat to Cherbourg, which was reached on 18 June. During the move 5th KOSB provided the divisional rearguard together with 554th Field Company Royal Engineers and they were the last British units to leave French soil. Before boarding HMS *Manxman* the rearguard had placed explosives on the main dockside facilities and as they left they watched the cranes blowing up one by one. It was a poignant moment. As the regiment's war historian put it: '"We'll be coming back!" shouted the Scots; or in other words "Ye've no' seen the last of my bonnet and me!"' (The latter reference is to Sir Walter Scott's song 'Bonnie Dundee' from his play *The Doom of Devorgil* (1830) written in honour of John Graham of Claverhouse, Viscount Dundee. Ironically, he was the commander of the Highland forces which fought against Leven's Regiment at Killiecrankie in 1689.)

1940–45: FRANCE AND NORTH-WEST EUROPE

1st, 4th, 5th and 6th battalions

At the time, the evacuation of the shattered British Army at Dunkirk was hailed as a 'miracle', but the new prime minister, Winston Churchill, got nearer to the truth when he said that 'never has a great nation been so naked before her foes'. As was the case in 1914, there was an immediate call for fresh soldiers but this time the bulk of them were provided not by volunteers, as had happened in 1914, but by the introduction of conscription. The National Service (Armed Forces) Act of September 1939 made all able-bodied men between the ages of 18 and 41 liable for service 'for the duration of the hostilities' and within three months 727,000 had registered for service. Subsequent legislation brought in a total of 4.32 million conscripts to serve in the armed forces between 1941 and 1945. To meet the influx the army had to change, and in August 1940 No. 10 Infantry Training Centre came into being at Berwick-on-Tweed to train recruits for The Royal Scots, The Royal Scots Fusiliers and the KOSB. For most of those recruits the years between the disaster at Dunkirk and redemption at D-Day in 1944 formed a time of retrenchment and retraining as

Britain struggled with the need to fight in North Africa and Italy as well as in Burma. Home defence duties were also paramount, and between 1940 and 1944 the KOSB's Regular 1st battalion and four Territorial battalions were involved in what the military historian John Keegan called 'endless divisional "schemes" and regimental manoeuvres, and fired their weapons at paper targets and squares on the map'. Not that it was without its dangers: while training at Epping 6th KOSB suffered 60 casualties (30 killed) when the Luftwaffe attacked a training position at Theydon Bois, where two platoons of A Company had been billeted. On a happier note the battalion beat 1st Welsh Guards 11–0 at rugby in the Northern Command Championship. (Of the original Borderers' XV five were killed and five were wounded in later operations.)

From the outset of US involvement in the Second World War, following the Japanese attack on Pearl Harbor in December 1941, American military planners had made a strong case for an early attack on the European mainland. In fact, the decision to press ahead with the invasion of north-west Europe had been taken as early as May 1943 at the Allied conference in Washington and under joint British–US direction planning for it began immediately after the summit had ended. By the end of the summer the plan was shown to the Allied leadership at the Quadrant conference in Quebec. The chosen landing ground was the Baie de la Seine in Normandy between Le Havre and the Cotentin peninsula, an area which met all the criteria, including a deepwater port at Cherbourg. It was agreed that the initial assault should be made by five divisions: two US, two British and one Canadian, with one British and two US airborne divisions operating on the flanks. The D-Day invasion, as it became known, began on 6 June with the airborne forces securing the flanks overnight while the main assault went in at dawn, preceded by a mighty bombardment from

2,000 warships in the Channel. By the end of the day the assault divisions were ashore and the five landing areas – Utah, Omaha, Gold, Sword and Juno – had been secured, with the loss of fewer than 10,000 casualties (killed, wounded or missing), not as many as expected.

Amongst the formations in the first phase of the invasion was 1st KOSB as part of 3rd Infantry Division, which landed with 50th (Tees and Tyne) Division and 3rd Canadian Division. Despite fears that 3rd Division would suffer heavy casualties as the spearhead at Sword beach, the landing was relatively trouble-free and by nightfall 1st KOSB was ashore and had assembled between Beauville and Benouville. For the next four weeks the fulcrum of the fighting in Normandy was the battle for Caen, which was eventually captured on 9 July. Also taking part in the operation was 6th KOSB, which had landed with 15th (Scottish) Division at the end of June and was quickly involved in the fighting. In one incident near Troarn the Germans kept firing until the Borderers were within yards of their position, when they raised their hands in surrender. 'Much good it did them,' was the caustic response of one of the officers, Major Charles Richardson. During the fighting – known as Operation Epsom – the battalion lost over 150 casualties. Robert Woollcombe's history of the 15th (Scottish) Division provides a telling description of the Borderers' reactions as they waited to experience their first shock of battle:

> We were in a fantastic world of unbelief. We felt nervously energetic. Officers and sergeants must preserve calm exteriors, appear pointedly casual; but the effort of shouting orders above the din spoiled the effect. The Jocks felt the moment. Some joked – goodness knows what about, but it didn't matter. Some stood silent, smiling apart as they listened to the enormous effort of the guns. All

> were joined by a most poignant undercurrent of emotion that obliterated rank. All were smoking. It was steadying to smoke.

The capture of Caen provided the Allies with the toe-hold that they needed to consolidate the D-Day landings and produced the impetus for Operation Cobra, mounted on 25 July by the US First Army, which pushed as far south as Avranches and the pivotal neighbouring town of Pontaubault. Suddenly the possibility opened of invading Brittany in the west and racing eastwards toward Le Mans and the River Seine. By the beginning of August, 1st KOSB was engaged in operations in the vicinity of Vire to the south-west of Caen, where the injunction to the advancing British columns was 'Drive slowly! Dust brings shells!' (The German artillery was quick to pounce on any evidence that Allied convoys were moving through the countryside.) The next problem for the Allies was overstretch – as the attacking forces moved away from the Channel beach-heads their supply lines became longer and that had an impact on the speed of the advance into north-west Europe. It also meant that the war would not end in 1944.

While these events were taking place the 4th and 5th battalions had been undergoing a bewildering series of changes of role. Following their brief foray into France with 52nd (Lowland) Division they had returned to Britain, where they were employed on home defence duties, first in eastern England and then at Kirkintilloch. This was followed by training in the north-east of Scotland, with the division in its new and specialised role as mountain warfare troops prior to an anticipated invasion of Norway. From autumn 1942 and throughout the following year the battalion was involved in a series of arduous training exercises in the Cairngorms which involved the men living in the hills for weeks at a time. Instruction was also given in fighting in snow conditions, skiing and handling of loads on horses and mules. The mood was best encapsulated in a rhyme written by Private R. McGill, 4th KOSB:

> Reveille at six and away up the hill
> With schemes night and day we were put through the mill,
> Till our legs and our feet were all swollen and sore
> Climbing heather-clad hills around Aviemore.

The culmination was training in combined operations at Inveraray but by then, the spring of 1944, it was clear that the invasion of Norway was no longer a priority. Instead, the division was given a new role as an air-transportable formation. The idea was to use the division in support of airborne operations by landing it with its own transport (jeeps and trailers) after parachute troops had secured the ground. It was an ambitious concept and a number of potential targets were identified including the Brest peninsula and the forest of Rambouillet south of Paris, but the speed of the Allied advance after D-Day put paid to any of the plans being put into effect. A more ambitious plan to use the 52nd (Lowland) Division as air-landing troops in support of the 1st Airborne's ill-fated operations at Arnhem also failed to materialise for reasons which are described at the end of the chapter.

Eventually, in the middle of October, the division was re-designated again, this time as an ordinary infantry formation when the 52nd (Lowland) Division was given the task of opening the port of Antwerp under the operational command of the First Canadian Army. And so it was that, having been trained for mountain warfare, 4th and 5th KOSB eventually went into battle below sea level. At that stage of the war the Allies had successfully completed the D-Day landings and broken out of Normandy. However, the speed of their advance had produced enormous problems of supply and re-supply – the further they moved into Europe the further away were the main Channel ports which provided them with much-needed fuel, munitions and the materiel of war. The one port capable of giving them everything they needed was Antwerp, with its huge docks, but it was still in German control and its seaward approaches were heavily mined. With winter approaching – the attack began on 31 October – the need to capture Antwerp and to open up the Scheldt estuary was imperative. The two KOSB battalions landed at Ostend with 7/9th Royal Scots and regrouped in the nearby town of Waereghem, where they were given a hero's welcome and offered tremendous hospitality. Their first objective was Flushing, in Walcheren, which the Germans had earmarked as the key to open the Scheldt estuary and had reinforced accordingly. To 5th KOSB fell the task of clearing the eastern part of the town as far as the Middleburg canal, and during the operation the men were under continuous German mortar fire. At the same time 4th KOSB supported 7/9th Royal Scots to capture the German headquarters in the Hotel Britannia. All the while, as the regimental war historian made clear, both battalions had to contend with atrocious conditions:

> Battle situations are frequently described as 'fluid'. Flushing literally was a fluid battle; it was aquatic in more senses than one. It began with an amphibious assault and, as it developed, the infantry often had to go into action waist deep in icy water. The waters of the Scheldt, pouring through the gaps in the sea wall, converted some of the roads into fast flowing rivers with treacherous currents at high tide and the men of the rifle companies had to hold on to each other in a human chain at some of the whirlpool spots.

In spite of the difficult conditions Flushing was in Allied hands by 3 November, allowing the Scheldt estuary to reopen to Allied shipping and the first transport ships were able to enter Antwerp by the end of November. During the operation 4th KOSB sustained 75 casualties, including three killed, and seven died of wounds, while 5th KOSB's losses were 62, nine of whom were killed.

The opening of the port of Antwerp was a vital ingredient in the Allies' move into north-west Europe. During this phase of the war 1st KOSB was based in the Venraij sector on the River Maas, 6th KOSB was still with 15th (Scottish) Division between Liesel and Meijil, while the 4th and 5th battalions were stretched out along an extended line between the villages of Birdgen and Kreuzrath. For all of the Borderers the weather conditions proved to be difficult and demanding – short, bitterly cold days with frozen fields and damp, comfortless trenches. At the same time the Germans put up a stout defence and there were also the added dangers of anti-personnel mines and other assorted booby traps. Then there were full-scale battles such as the attack on Blerick on 3 December, which involved 6th KOSB. The town was home to the main Eindhoven to Cologne railway line and was the last German stronghold west of the River Maas. As a vital strategic asset it was well defended, but meticulous planning and the use of overwhelming armoured support carried the day. So successful was the operation that within the 15th (Scottish) Division the town gave its name to new tactics, a 'Blerick' attack being used in similar assaults on heavily defended positions.

Still in the Geilenkirchen sector, 4th and 5th KOSB opened their account in 1945 by taking part in Operation Blackcock, an offensive against German positions in the triangle of land between the rivers Maas and Roer. This was a large-scale operation undertaken in bitterly cold conditions with infantry and armour combining in a nine-day battle which fully tested those who took part in it. Thanks to their training as mountain troops, the men of 52nd (Lowland) Division were properly equipped – as the regimental war historian explained, 'at last the "Lowland Mountaineers" were in the arctic warfare for which they had been trained so thoroughly in the Scottish Highlands'. In that role both the KOSB battalions acted in combination at Waldfeucht, the first German town to be fought over by British troops in 1945, and one which was generally held to be the scene of the most serious fighting during Operation Blackcock. This was a fiercely contested battle in which infantry had to take on and destroy German tanks, including the Tiger with its much-feared 88mm gun, capable of piercing armour at a range of 2,000 metres. This was no easy task, given the fact that the men were operating in a built-up area. At one stage in the fighting a section of D Company 5th KOSB engaged a Mark IV tank with a PIAT weapon (Projectile Infantry Anti-Tank) only to see the shell bounce off the tank and rebound into the room of the house from which it had been fired. The successful outcome of the fighting in the Roer triangle proved to be a decisive move in the operations to break the German defensive positions on the West Wall.

Ahead lay the final operations of the war as the Allied forces made ready to push from the Maas towards the Rhine and then move into the German heartlands. This period provided 6th KOSB with one of its last set-piece battles of the war, the struggle for the town of Goch, which was very much a Scottish battlefield, involving as it did 15th (Scottish) Division and 51st (Highland) Division as well as 1st Highland Light Infantry from 53rd (Welsh) Division. During the fighting, which lasted 18 days, the battalion lost six killed and 29 wounded. While this was happening – the assault was listed as Operation Veritable – 4th and 5th KOSB advanced towards Gennep and Afferden and again met fierce resistance in the shape of heavy mortar and machine-gun fire. At the end of February 1st KOSB took part in Operation Heather, which secured the Uden–Weeze road. During the fighting the men rode into battle on board 50 'kangaroos', specially adapted Sherman tanks with their gun turrets removed, and escorted by Churchill tanks and self-propelled artillery. (The kangaroos were prototype armoured personnel carriers and had no heating; Brigadier Frank Coutts, a distinguished Borderer and at the time a temporary captain in charge of 4th KOSB's anti-tank platoon, remembered men huddling around the exhaust trying to get warm when the machines halted.) In one incident six stretcher-bearers from the 5th battalion, all bandsmen, were killed when their dug-out received a direct hit. As the month came to an end the Germans had been pushed back from the Maas and the Allied armies – Ninth US Army, Second British Army and First Canadian Army – were ranged along the west bank of the Rhine prior to crossing it and then wrapping up the industrial conglomerations of the Ruhr.

On 23–24 March 6th KOSB had the distinction of being one of the first formations to cross the Rhine in amphibious Buffalo vehicles, the task of 44 Brigade being to link up with the airborne forces which had been dropped to the east of the river. Once again the Germans offered fierce resistance, but Germany's western frontier was finally breached and the end of the war was in sight. As the commander of 44 Brigade put it, the day had been an unalloyed success: 'The day ended triumphantly. The Lowlanders had stormed the Rhine, captured a thousand German prisoners and many guns, relieved the airborne troops, and opened up the way for the breakthrough. The casualties had not been light, but the soldiers were in tremendous spirits.' The divisional historian was more prosaic, describing the atmosphere in the assault battalions as 'like being in the dressing-room at Murrayfield or Hampden before an international match'. While apposite for a regiment which recruited from the heartlands of Scotland's rugby community – at the time the Borders provided large numbers of players for the national rugby XV – the sporting analogy did not tell the whole story. As the author of the regiment's war history, Captain Hugh Gunning, made clear in his description of the action, the crossing of the Rhine was more than a military action, it was also a declaration of intent:

> The Rhine is the proudest boast of the German. The Glaswegian sings fondly of the bonnie banks o' Clyde, but it is a homely song. The American sings of the Mississippi, but it is a genial folk song of cotton fields and simple things. The Egyptian sings of the Nile, and the Russians of the Volga, but their song is over river-worship of a restrained kind by comparison with the song the German sings of Father Rhine. The Rhine to the German is a mystical river which rises in the sources of the most extravagant nationalistic fantasy in Europe. The Rhine is the river of the German Valhalla. The German gods are Rhine gods, his maidens are Rhine maidens, his heroes Siegfried and the others of the German Operatic Company, are Rhine heroes.

Gunning was writing at a time when the Nazis' fascination with Germany's Wagnerian past was a main driver of Hitler's rise to power, but he understood exactly the symbolic importance of the Rhine crossing. Now the Allies were in the German heartlands which Hitler and his cohorts had always said would never be breached. From the Rhine the Borderers moved towards Hanover and another river crossing, the Elbe, on 29 April. A week later came news of the ceasefire and the end of the war in Europe.

In those last days of the war 1st KOSB crossed the Rhine at Rees on 27 March and although it met sustained opposition at Lingen, where the garrison was manned by Hitler Youth soldiers, much of its work was mopping up German positions. In a sharp action at Bramsche the 1st battalion surprised and overwhelmed a German anti-aircraft regiment. Following in its wake was 4th KOSB, which crossed the Rhine on 28 March to the strains of a hunting horn sounding the chase. Now under the command of 7th Armoured Division, the famed 'Desert Rats', the battalion encountered stiff resistance while moving through the Teutoberger Wald. Again, much of the opposition was provided by fanatical boy-soldiers who were often prepared to fight to the last man. Also involved in the chase into Germany was 5th KOSB riding on 'kangaroos' alongside tanks of 4 Armoured Brigade. Its last serious piece of action came at Voltlage on 8 April, where the opposition was provided by NCOs from a nearby army training school. In his memoir for the BBC James MacQuarrie described German soldiers refusing to surrender even when flame-throwers were used against their positions: 'A lot of men came out on fire but they were still armed and firing at us, so we had no choice but to fire back.' During the action the battalion lost one officer and 15 soldiers killed.

For 6th KOSB there was one final battle, at Uelzen on 15 April, during the advance towards the Elbe, where the town was quickly captured and then held against determined German counter-attacks. Bremen was the final goal for the 1st, 4th and 5th battalions, while 6th KOSB crossed the Elbe and headed north towards the Hanseatic port of Lübeck, which had been spoken of as a possible centre for German resistance (the threat never materialised). Then it was all over bar the need to pacify the local population, deal with the huge numbers of displaced persons who were on the move throughout Europe and to confront the terrible evidence of the concentration camps. As Lieutenant Peter White reflected on the moment in his memoir *With the Jocks* – one of the best soldier's accounts of the conflict – the end of the war was met with the same restraint that had greeted the armistice 27 years earlier (see Chapter Six): 'I had never pictured that the reception of this so long hoped for and magical news of the war's end would have been taken so quietly . . . later that night the sounds of a sing-song, no doubt helped by the rum issue, broke out from one platoon, but that was all. Life went on just as before.'

1944: ARNHEM

7th battalion

As the international situation deteriorated in the 1930s the decision was taken to expand the Territorial Army and early in 1939 the 7th (Galloway) battalion came into being, formed from elements of the 5th battalion. Its first commanding officer was Lieutenant-Colonel The Earl of Galloway. As was the case with other similar infantry units, it was mobilised on 27 August as a home defence battalion and remained in Scotland until May 1940 when it moved south, first to Marlborough and then to Byfleet in Surrey. On taking up its place in the line it formed part of 44 Brigade, with 6th KOSB in the 15th (Scottish) Division. Initially its duties were confined to home defence and to providing drafts for other formations fighting in North Africa and it was not until 1942 that the raw recruits became efficient fighting soldiers, ready for battle. In October 1942 the 7th battalion was moved north to Orkney and Shetland on coastal defence duties, but this was only a prelude to a more startling change of direction. As part of 1st Air Landing Brigade the battalion began the transformation to airborne soldiering under the command of Lieutenant-Colonel Robert Payton Reid. The battalion's next destination was Woodhall Spa in Lincolnshire, where the brigade began training in earnest for its new role as airborne troops who would be taken into action by glider. Formed in 1941, the 1st Airborne Division consisted of two parachute brigades and one air landing brigade and it had already seen action in North Africa and Sicily. Some idea of the intensity of the training can be found in the diaries of Albert Blockwell, an English-born Borderer who had started his war in the Royal Army Ordnance Corps before transferring to 7th KOSB:

> We began to get fully equipped with new weapons, new trucks, etc., and we were kept busy for weeks travelling to different parts of the country for trucks, motor-bikes etc. We collected 70 lightweight motorbikes from Peterborough and 70 heavy motorbikes from Birmingham. In the meantime, drivers collected trucks and Jeeps from other places, so finally, after a few weeks, we were up to full strength in transport. Eventually we got our full equipment too, radio sets, small arms, medical supplies and extra equipment that we had to carry.

Airborne soldiering was not just about hard training with new portable equipment: it also allowed the battalion to exchange their bonnets for the coveted red beret and Pegasus badge of the airborne forces. For 7th KOSB this meant getting used to the idea of going into action on board a 26-seat Horsa glider which also carried a handcart with the platoon's ammunition reserve and a small motor-cycle, and was generally towed towards the dropping zone by a converted bomber-tug such as a Stirling or Albemarle. (During the operation a larger Hamilcar glider was also used.) There was also an expansion in size; together with the other battalions in the brigade (1st Border and 2nd South Staffords) 7th KOSB consisted of four rifle companies, giving a total of 16 rifle platoons, each one carried in a Horsa glider. While there was no carrier platoon, the 3-inch mortar, machine-gun and 6-pounder anti-tank platoons were doubled in size to provide additional firepower once on the ground. Unfortunately a disaster befell the battalion during an exercise on 4 April 1944 when a Stirling tug towing a Horsa carrying No. 3 Platoon hit a tree on high ground and crashed, killing 26 Borderers together with six aircrew and the two glider pilots.

The final objective of the preparation was an audacious plan to use airborne forces (codenamed Market-Garden) during the operations in the Netherlands following the D-Day landings of June 1944 and the subsequent breakout from Normandy. It was designed to outflank the German defensive positions known as the West Wall by establishing a bridgehead across the lower Rhine at the Dutch town of Arnhem and thereby threatening the approaches to the Ruhr. This would involve the capture of the bridges over the Maas at Grave, over the Waal at Nijmegen and the Lower Rhine at Arnhem. There would be two phases: Market, the operation to seize eight river-crossings using British and US airborne forces and Garden, the advance of strong armoured forces to support them from a start-line on the Meuse–Escaut canal 59 miles away. It was a bold concept but it called for accuracy in the drop zones and speed on the part of the ground forces to give quick support to the airborne units. In the British division there were two parachute brigades and an air-landing brigade backed up by the Polish Independent Parachute Brigade, all under the command of Major-General Roy Urquhart, formerly of The Highland Light Infantry. Overall command of the airborne operation was in the hands of Lieutenant-General F. A.M. 'Boy' Browning (husband of the novelist Daphne du Maurier).

Alas, the operation was a failure. Although the British and US airborne forces completed their drops on 17 September with unexpected accuracy – US 82nd Airborne Division at Grave and Groesbeek, US 101st Airborne Division at Eindhoven and Vegel, British 1st Airborne Division at Arnhem – there was a steady accumulation of delays which held up British XXX Corps. Added to stout German resistance – unknown to the Allies two SS Panzer divisions were refitting in the area – the airborne forces were up against it from the outset. That much became clear when 7th KOSB landed in the first wave and its pipers began playing 'Blue Bonnet O'er the Border' to rally the men into their company positions. One Horsa landed in the sea and its occupants, under the command of Lieutenant W. G. Beddowe, became the first prisoners of war of the Arnhem operation. The rest of the battalion moved westwards to secure the drop zone of the following day instead of proceeding to Arnhem. There were also a number of crashes, one of which was described by Sergeant George Barton in Martin Middlebrook's account of the battle:

> Everything happened in a rush. We were in a dive, then the glider suddenly lifted and the undercarriage caught in the trees. I think the pilot was trying to get over the trees, but everything came to a dead stop. I was flung against the front of the jeep, and my equipment all broke away from me. Fortunately, the jeep was well fastened down, and the gun didn't move either. The glider was nose into the trees with its tail in the air.

Ahead lay over a week of heavy fighting against the German defenders, who had in fact been training for operations against airborne landings. An indication of the problems facing the battalion came on the second day, when D Company came under attack on the southern side of Ginkel Heath, the dropping zone for 4 Parachute Brigade. By contrast, B Company destroyed a number of German half-tracks and killed their Dutch SS crews. During the operation to hold the heath for the actual drop 7th KOSB lost two officers and 13 men. The battalion's next task was to guard the ground near Johannahoeve Farm for the next landing of glider-borne forces and once again they came under determined attack by enemy forces in the heavily wooded area; during the fighting A Company was overwhelmed and forced to surrender.

By then the battle was slipping away from the British airborne forces. The two parachute brigades had failed to reach their objectives and as a result of accidents and enemy action most of the Polish Brigade had failed to arrive or were badly shot up as they landed under heavy enemy fire. The following day B Company was overwhelmed and forced to surrender and the final act took place within the Oosterbeek perimeter, where the battle continued until 25 September. It was known to the Borderers as the Battle of the White House after the Dreyeroord Hotel, which stood on the north-east corner of the perimeter and is still in existence. All told, 765 Borderers went into battle and of those 112 were killed, 76 were evacuated and 577 were listed as 'missing', most of them becoming prisoners of war. Colonel Payton Reid was the only British battalion commander to be evacuated after the survivors succeeded in crossing the Lower Rhine and fighting their way to Nijmegen. Only 2,163 members of 1st Airborne Division succeeded in making their way back to the Allied lines. Most of the Borderers' casualties are buried in Arnhem Oosterbeek Cemetery. Amongst their number is the grave of Major Edward Coke, 6th KOSB, who was killed with the ground forces nearby and whose body was brought to Arnhem to lie beside his brother, Major John Coke, 7th KOSB, who survived the battle but was killed while leading a party of evaders during the final break-out.

It was not the end of 7th KOSB's adventures. Following the collapse of Germany in May 1945 the battalion flew with 1st Airlanding Brigade to accept the surrender of the German forces in Norway.

In severe weather conditions one of the battalion's Stirling tugs crashed, killing four men. After returning to Scotland the battalion was disbanded on 28 November 1945 and its men were posted to other battalions including 2nd KOSB and 4th Black Watch.

NINE

The Second World War: India and Burma

Like the earlier conflict, the Second World War was fought on a global scale. Britain and its allies were also engaged in a punishing and increasingly brutal round of fighting against Imperial Japan, whose expansionist policies in South-East Asia and the Pacific had drawn it into the conflict in December 1941. For Britain and the US the war against Japan had begun badly. In an act of infamy seldom before equalled in history, Japanese naval and air forces had unexpectedly attacked the main US naval base at Hawaii, Pearl Harbor, on 10 December 1941 and destroyed a substantial part of the US Pacific fleet. Britain's crown colony, Hong Kong, had fallen on Christmas Day 1941 and this was followed by the rapid collapse of the British defences in Malaya. Although outnumbered two to one, the Japanese army had fought with an aggression and *élan* which had confounded British commanders and overwhelmed the inadequate British, Australian and Indian defence forces. On 15 February 1942 Allied fortunes reached a nadir when Singapore, with its huge naval base, capitulated to General Tomoyuku

Yamashita's Twenty-fifth Army and 130,000 Allied soldiers went into captivity.

Equally disastrous was the retreat from Burma, which followed at the beginning of 1942 when it was invaded by the Japanese from Raheng in Thailand. Originally the Japanese had not been interested in occupying the whole country and believed that their strategic needs would be served by taking the port of Rangoon and the airfields on the Kra isthmus, but their minds were changed by the realisation that Britain could use Burma as a springboard to attempt to retake Malaya, and also by the threatening presence of the Chinese Fifth and Sixth Armies to the north along the lines of communication known as the Burma Road.

The Japanese plan to rectify the situation was based on a three-pronged attack – on Rangoon, the Salween River and the Sittang River – and as in Malaya they relied on speed and aggression to accomplish their objectives. On 11 February they crossed the Salween; the retreating 17th Indian Division blew up the bridges across the Sittang three days later and by 18 March 1942 Rangoon had fallen. Although the British and Indian forces counter-attacked in the Irrawaddy Valley at the end of the month, they were outflanked to the east and west, where the Japanese drove General Chiang Kai-shek's army back towards the Chinese border. Short of supplies, exhausted and demoralised, the two armies went their separate ways and the British and Indian forces began what came to be known as 'the longest retreat in British military history'. Following a march of 900 miles, the survivors crossed over the border into India on 19 May: of the original 30,000, 4,000 were dead and another 9,000 were missing.

In July 1942 the Japanese high command made plans for 'Operation 21', another three-pronged attack from within Burma towards Ledo, Imphal and Chittagong in the north. It was over-ambitious, as the terrain in northern Burma was not suited to the rapid offensive operations which had been used so successfully in the Malay peninsula, but the fact that India was threatened was enough to convince British minds that their position was precarious. At the same time, Indian nationalists were making increasingly strident demands for Britain to quit India and there was a clear and immediate need to restore British prestige by taking the offensive back to the Japanese and retrieving lost ground in Burma. The opening initiative was the first Arakan campaign, which began in September 1942 and was aimed at capturing the Akyab peninsula following an advance from Chittagong by way of Cox's Bazaar and Donbaik. Some ground was won, but by the following May the Japanese had retrieved it all; for the British it was not only an expensive failure which cost over 5,000 casualties but it inculcated a belief that the Japanese were unbeatable jungle fighters.

It was against that background that the first useful steps were taken to retrieve the position. For the first time, specially trained British and Indian soldiers showed that they were capable of taking on and beating Japanese soldiers in the fastnesses of the Burmese jungle. The creator of the turnaround was Major-General Orde Wingate, a remarkable gunner officer with unorthodox opinions, who had served in Palestine before the war and had earlier helped the Emperor Haile Selassie to return to his throne in Ethiopia. He believed that the Japanese could be overcome by inserting long-range penetration forces which would operate behind enemy lines and destroy vital objectives. Fighting in eight columns supplied from the air, the force was called the Chindits (after the Burmese word *chinthe* the mythical winged beasts which guard Buddhist temples) and it went into action in February 1943. Operation Longcloth (as it was known) was a mixed success. It sowed confusion in the minds of the Japanese high command, who feared it was a precursor to a large attack and so tied up troops to hunt down the Chindits. The

Mandalay–Myitkyina railway line was cut but the cost was appalling. Of the 3,000 men who carried out the operation, only 2,182 came back; around 450 had been killed in action and the remainder were either lost or had been taken prisoner. Only 600 of the force were able to return to soldiering. Nevertheless, Wingate had shown that the Japanese could be fought on their own terms and the Chindits were a huge propaganda success, so much so that a second, larger operation was planned for the spring of 1944.

1942: INDIA

2nd battalion

Prior to the outbreak of the war with Japan the 2nd battalion had been serving at Razmak in Waziristan, a tribal area in the North-West Frontier Province, which had spent five years in a state of almost perpetual warfare involving Mirza Ali Khan, the Faqir of Ipi. With the coming of war, fears that either the Germans or the Russians would take advantage of perceived British weaknesses were not exaggerated. For long an implacable enemy, the Faqir of Ipi was in receipt of German and Italian funds and he was opposed to the construction of anti-tank defences and other measures in Waziristan. The situation was not helped by the reverses suffered by the British Army elsewhere: every Pathan knew that the Germans had driven the British out of France, Greece and Crete and that the Japanese had inflicted humiliating defeats in Burma and Malaya. All this added to the Faqir of Ipi's prestige and the security forces were hard pushed to keep the peace when his forces attempted to capture the fort at Datta Khel, the most exposed of the frontier garrisons. Together with the whole brigade (1st Queen's Own Royal Regiment and 1st Somerset Light Infantry), 2nd KOSB moved on the besieged fort which was in fact well supplied with food and ammunition, contributing to a successful outcome for the defenders. During the operations at Datta Khel 2nd KOSB lost the adjutant Captain John Agar and Lance-Corporal Halliday, as well as nine soldiers who were wounded. The Faqir's failure to take Datta Khel was a blow to his prestige, and the arrival of better news for the British from the main battlefronts encouraged wavering tribal chiefs to hedge their bets – as winter began to set in news arrived of the Allied victory at Alamein, the German defeat at Stalingrad and the Japanese failure to invade India.

1943–45: BURMA

2nd battalion

Following their deployment in Razmak, which lasted over a year, 2nd KOSB joined the Fourteenth Army in August 1943 for the campaign in Burma. The battalion's eventual destination was the Arakan peninsula, but before then it trained between Peshawar and Rawalpindi with 89 Brigade in Major-General Frank Messervy's 7th Indian Division. The other battalions in the brigade were 7/2nd Punjab Regiment and 4/8th Gurkha Rifles. This was followed by specialist jungle training at Singhori, near Chindwara in the Central Indian Province (today Madhya Pradesh), where it was joined by training teams made up of soldiers who already had experience of the vastly different conditions of fighting in Burma. According to the regimental war historian, it was a steep learning curve for all the Borderers:

> These courses of instruction, ranging from divisional level down to companies and platoons, gave every man in the division a chance to learn quickly and thoroughly all that could be taught to make him fit for battle in the jungle. The individual training was carried out in company jungle camps where the men spent five days, returning to Singhori for two days and then back to the jungle camp. The Borderers

> were shown how to use bamboo, how to navigate in dense jungle and elephant grasses, and how to live on the land; in brief, how to conquer the jungle as a necessary preliminary to beating the human enemy, the Japs.

This was very much the credo adopted by Wingate's Chindits: the jungle need not be a hostile place, approached differently it could be an environment which offered a measure of protection. And above all, the Japanese opposition were not supermen but ordinary soldiers who were well capable of being beaten. Training also included live firing exercises and other toughening-up exercises which were carried out in all weather conditions to make sure that the troops were readied for whatever the shock of battle would throw against them. The chance to put theory into action came at the end of August 1943, when 2nd KOSB moved by train from Ranchi to Madras, where they boarded the troopship *Ethiopia* for a four-day voyage across the Bay of Bengal to land in Chittagong on 20 September 1943.

This was the southern front of the Allied operations against the Japanese, the other two being the central front, with its main battlefields at Imphal and Kohima and the northern front bounded by Yunnan and Ledo. Morale was low following the setbacks in Malaya and Burma and the failure of the first Arakan campaign in September 1942–May 1943. However, by the time the Borderers reached the Arakan in August, the monsoon had arrived and both sides were more than content to sit it out until the weather improved. As a result 2nd KOSB saw little of the enemy as it settled into its positions between the Mayu Hills and the Naaf River, which was described as 'a waste of flooded fields with hillocks covered in jungle scrub'. Although the hills were not particularly high, they had steep precipices and the ravines were deep and threatening, with their fair share of insects and snakes. As was the case on other fronts in Burma, rations and ammunition had to be carried in by mule trains.

As the weather began to improve, 2nd KOSB was able to send out fighting patrols to engage Japanese ration parties and scored a first success on 8 October when an ambush succeeded in killing 30 of the enemy who came from an Imperial Guards regiment. This was followed by a similar encounter a few days later on a position known as the 'Horseshoe', in which the Japanese had unwisely advertised their presence by flying their national flag on the high ground. It soon became apparent that casualties would not just be caused by enemy fire, sickness too was a problem. In the first months in the Arakan the battalion's casualty rate was 120 sick men (mainly malaria and dysentery) to every man killed or wounded by enemy action. Gradually better standards of hygiene were introduced and by 1944 the ratio had fallen to six to one.

Following a period of recuperation, the battalion moved south by way of the Ngakyedauk Pass (known throughout the Fourteenth Army as 'Okedoke' Pass) to new positions opposite the Japanese lines at Tatmin Chaung. The aim of the commander of XV Indian Corps, Lieutenant-General Sir Philip Christison, was to recapture Akyab so that its vital airfields could be used to operate against Rangoon. This phase of the operations provided the Borderers with their first set-piece battle against the enemy, but it was preceded by a lengthy game of cat-and-mouse. At that stage of the fighting the British and Indian armies in Burma lacked detailed intelligence about their opponents and a high premium was placed on reconnaissance patrolling, with the objective of bringing in a live prisoner. (To add a sense of competitiveness to the process there was a reward of 250 rupees and 28 days leave for anyone who succeeded in doing so.) Although 2nd KOSB failed on that score – a raiding party did bring in one prisoner but he was found to be dead on arrival following a blow to the head as he tried to escape – the deployment did bring a successful engagement with the Japanese at the beginning of 1944. This took place on a position known as Able Hill, where the objective was to cut the Japanese lines of communication with Mungdaw and it was conceived, in its first stages, as a night operation. With 1st Queens on the right and 4/5th Gurkhas on the left, 2nd KOSB attacked in the centre towards Ledwedet Hill and immediately ran into fierce enemy fire which killed the commanding officer Lieutenant-Colonel W. G. Mattingly (known throughout the battalion as 'Matt'). After two weeks in the line, the battalion was relieved by the Gurkhas and returned to the start line, where the records show that the men were treated to comforts such as warm food, clean clothes and blankets, and rum and cigarettes.

After a short period of rest the battalion moved to a new position in Wet Valley, where it relieved 1st Queens and then prepared to move north again towards Taung Bazaar, which had been occupied by 4,000 Japanese troops. This was a difficult and demanding operation, which had to be carried out in unknown territory and as the regimental war history explains, it involved a night march, something that most soldiers fear and dislike:

> There were no guides, no maps, and this stretch of the country was unfamiliar. A course was set on a compass bearing and the column set out. A mule column is not easily kept quiet, but the animals seemed to sense the danger. The eerie march in the misty moonlight took four hours, and the column was duly navigated to the rendezvous. The mules by a mischance had acted as the spearhead of the brigade on this move, but the column had luck on its side.

On arrival at its new position at Allwynbin, four miles south of the Ngakyedauk Pass, 89 Brigade constructed its 'admin box', the main centre for communication (in reality shaped more like a bowl than a box), in the lee of a position called Sugar Loaf Hill. There the brigade was joined by 33 Brigade and 114 Brigade, and all three were now fenced in by the opposition. Air supply became imperative and here the brigades were helped by the fact that the RAF enjoyed air superiority and its Spitfires were more than capable of shepherding Dakota transports full of supplies to their targets. The arrival of food, cigarettes and other comforts came as a much-needed relief to the men of 2nd KOSB, who had gone five days without a square meal. Expectancy turned to disappointment in one instance where the containers held ammunition, but this was balanced by others which were replete with 'food and smokes'. This provision from the skies was a real morale-booster while fighting in difficult terrain and facing an unyielding enemy. Not only did it bring much-needed provisions to the beleaguered men in the admin box but it produced solid evidence that others knew about their plight and were doing their best to help them. Bucked up by that kind of support, battalions like 2nd KOSB always rediscovered the urge to take the fight back to the enemy. Despite the difficult conditions, morale rarely slumped and in the pages of the SEAC (South-East Asia Command) newsletter there were glimpses of the kind of humour expressed by an anonymous Borderer about fighting in the Arakan (to be sung to the tune of a popular, if vulgar, rugby club song):

> Japs on the hilltops
> Japs in the Chaung
> Japs in the Ngakyedauk
> Japs in the Taung
> Japs with their L of C [lines of communication] far too
> long
> As they revel in the joys of infiltration . . .

Resupplied and suitably refreshed, 89 Brigade took the initiative and reopened the battle for the defence of the admin box in the first week of February. It quickly proved to be an intensive period of combat, often at close quarters and with mercy neither given nor expected. At one stage the Japanese broke into the medical dressing station and massacred those being treated. The nights were made hideous by howls and shrieks from the attacking Japanese, but despite the determination of their assaults the enemy attack was soon running behind timetable. The admin box remained secure, Ngakyedauk Pass was reopened, with 2nd KOSB in the vanguard, and after a grim 18 days of heavy fighting the British and Indian positions in the Arakan were secured. For the Borderers there was one more action on the Maundaw–Buthidang Road, where a massive artillery barrage preceded their attack on Japanese positions at Tatmin Chaung. In one chilling incident on the Horseshoe position the Japanese were seen to be pegging out a body in the hot sun as a lure. Although 2nd KOSB suspected that the unfortunate victim was a Borderer the position could not be attacked until nightfall, by which time the body had disappeared. During the mopping-up operations a position was found with 126 dead Japanese soldiers, all of whom had been ordered to commit suicide after losing the battle. On 26 February the Japanese called off their attacks in the Arakan: for the first time in the war British and Indian troops had managed to stave off a major Japanese offensive and the commander of the Fourteenth Army, General Sir William Slim, was suitably effusive in his appreciation of the British and Indian battalions which had brought the fighting in the 'admin box' to a victorious conclusion:

> British and Indian soldiers had proved themselves, man for man, the masters of the best the Japanese could bring against them . . . It was a victory about which there could be

> no argument and its effect, not only on the troops engaged, but on the whole XIV Army, was immense.

Following the failure of their Arakan offensive, the Japanese turned their attention to the north, attacking through Tiddim towards the Imphal Plain at Torbung and then through Tamu and Palel to cut the Allied supply road from Dimapur to Imphal, the most likely target being Kohima. The attack began on 14 March, spearheaded by the Japanese 33rd Division in the south and 31st Division further north with 15th Division in support, and Imphal was immediately put under siege. To meet the emergency, 2nd KOSB was airlifted in April from Sylhet to the area of operations, where it came under the command of 5th Indian Division. Almost immediately the battalion was given the task of holding a section of the Imphal–Kohima Road in the area of the Kanglatongbi Ridge. Despite the onset of the monsoon, engagement with the enemy was almost continuous and 2nd KOSB was involved in two major brigade attacks. In the first, against a formidable position known as the Hump, 2nd KOSB's pipers took the lead with the regimental charge, but the rifle companies were soon pinned down by withering machine-gun fire. To make matters worse, the Japanese defenders started rolling grenades down the steep hillside. Both the Gurkhas and the Sikhs were similarly discomfited and after three days of sustained action the Japanese were still in their original defensive positions. With stalemate approaching, 2nd KOSB was moved to higher ground south of the Hump where mules and porters brought in their supplies. It was not until 1 July that the rest of 89 Brigade was able to move off the Imphal–Kohima Road and push eastwards. As the regimental war history makes clear, it was tough going:

> The mountainsides were matted with jungle and scrub, and the valleys had a carpet of elephant grass 15 to 20 feet high.

> The troops had to hack their way through these grasses when they moved into the valleys which were broken by narrow streams, turbulent and deep. The villages perched on the hill tops, like keeps or forts, completed the panorama.
>
> The Borderers' march was an advance across the grain of the country; climbing the ridges and dropping down into the valleys thick in the elephant grass which harboured the dreaded scrub typhus bug. Officers and men carried all their kit on their backs, including their blankets, and the whole was sodden by the persistent rain.

Malaria proved to be as deadly as the Japanese, who had the unpleasant habit of booby-trapping the bodies of dead Borderers. On the other hand, 2nd KOSB was given invaluable assistance by the local hill tribesmen, who were no friends of the Japanese invaders.

A second brigade attack was mounted against a Japanese fortified position at Ukhrul in the middle of July. This too proved to be a perilous business as the Japanese were well dug in and enjoyed open lines of fire. Bad weather conditions prevented air strikes and it was not until reinforcements arrived that the Japanese decided to pull out as the 23rd Indian Division began its breakout from Imphal. By then 2nd KOSB was in a parlous state, with its numbers down to 250 largely as a result of malaria and a disastrous epidemic of scrub typhoid. On 25 July the battalion was pulled out of the line and sent to the Naga village of Nerhima in the vicinity of Kohima, where it went into a period of much-needed rest and recuperation. Even as late as August it was far below its war establishment: 25 officers instead of 32; 597 other ranks instead of the normal 816. However, the arrival of a new commanding officer, Lieutenant-Colonel H.R.R. Conder, and the appearance of a draft of 260 recruits quickly changed the situation. It says much for the new arrivals that although they were not Borderers – Conder had been commissioned in The

Royal Norfolk Regiment and many of the draft were from The Royal Inniskilling Fusiliers – they quickly adopted the KOSB ethos and in time became enthusiastic Scots.

The Borderers' experience helps to explain one of the main reasons for the Allied success at Imphal and Kohima. Whereas the Japanese had planned for a short, sharp campaign, the British and Indian divisions were there for the long haul. As casualties started mounting on both sides the Allies were able to re-equip and reinforce their front-line troops but the Japanese were denied that luxury. It helped that the RAF had command of the skies, allowing non-stop air drops to take place, but in stark contrast the retreating Japanese forces were on their own. By 19 August the River Chindwin had been re-crossed and British intelligence reports (some of which appeared in the *Official History* of the campaign) hinted at the plight facing the once all-victorious enemy:

> The Japanese, that legendary soldier who was supposed to live on a handful of rice from a little ration bag hung round his neck, began in fact to have no other ration. The effect was appalling, for even Japanese cannot live by rice alone. Fourteenth Army doctors reported that many who fell into our hands – desertion and surrender multiplied their number tenfold – were suffering from acute beri-beri. Their body cells had lost all power to absorb water. Their skins, stretched as taut as drums across the bone framework, were covered with dermatitis sores . . . And, ironically, many died of starvation with full ration bags of rice around their emaciated necks – rice which they could not eat.

By the end of December the retreat from Burma in 1942 was but a memory and Slim's Fourteenth Army had begun an offensive which would take it down the Irrawaddy line towards Rangoon by way of Mandalay and Meiktila, the twin keys to retaining possession of central Burma. A rejuvenated 2nd KOSB was part of that inexorable process, having rejoined the line in the Tillim area at the beginning of January 1945. Ahead lay days of lengthy marches, with frequent skirmishing. In the first week of February Pakkoku fell, paving the way for the crossing of the Irrawaddy. During this phase of the operations in its drive south, 7th Indian Division received armoured support from 116th Regiment (Gordon Highlanders); at the beginning of March Meiktila, with its rail head and two airfields, fell into the hands of the advancing XXXIII Corps commanded by Lieutenant-General Montagu Stopford. Since the advance from Kohima, 800 miles had been logged and now the Fourteenth Army was on its last lap. The final action for 2nd KOSB took place on the Prome Road at the end of May where the battalion lost seven killed and six wounded but succeeded in killing 141 Japanese. Following the fall of Rangoon – which was hastened by amphibious landings by 26th Indian Division – the battalion remained in Burma until 2 August, when it embarked on the *Egra* and moved back to Calcutta. Its casualties, recorded on the war memorial in Rangoon Cathedral, were 12 officers and 186 other ranks killed on active service. The total number of Borderers killed on all fronts during the Second World War was 82 officers and 1,269 other ranks.

The reconquest of Burma is one of the great sagas in the histories of the British and Indian armies. It was the longest sustained campaign of the Second World War; it was fought over a harsh terrain which included deep jungle as well as desert and mountains; it was often war to the knife with opposing soldiers caught in bitter close-quarter combat, and those who surrendered were rarely granted mercy. It began with a painful retreat and ended with a famous victory which relied as much on the endurance and fortitude of Allied troops as it did on the skill of their commanders.

It involved soldiers from a wide range of countries – Britain, India, Burma, China, Nepal, the United States and the West African countries of Nigeria, Sierra Leone and Gold Coast (later Ghana) – and because the campaign was almost as long as the war itself, it saw the introduction of innovations such as the use of air power in support of ground troops, modern radios to guide the strike and supply aircraft to reach their targets.

TEN

The Cold War, The Korean War and the Gulf Wars

Britain ended the Second World War victorious, but being on the winning side came at a high price and Clement Attlee's post-war Labour government found itself having to grapple with the problems of recession, shortages and financial restrictions imposed by the shattered economy. As a result, and as had happened so often in the past, shortage of funds meant that there had to be economies, and that meant cutting back the budgets of the three armed services. To some extent many things remained the same – despite losing their 2nd battalions in 1948, most infantry regiments remained intact – but the next 20 years were to witness a sea-change as the size of the army was gradually decreased and Britain's overseas colonial holdings were dramatically reduced. In a world which saw Britain negotiating a loan of $3.75 billion from the United States, followed by harsh measures to restrict domestic expenditure, huge armed forces were a luxury the country could ill afford. Between 1946 and 1948 the RAF Estimates shrank from £255.5 million to £173 million. The Naval Estimates for 1949 totalled £153 million, a decrease on the previous year's expenditure of £44 million and on both services the government urged further economies in personnel and materiel. Expenditure on the army was also reduced, from £350 million to £270 million, and Second World War equipment was not replaced in any quantity until the 1950s, forcing Field Marshal Viscount Montgomery of Alamein, Chief of the Imperial General Staff between 1946 and 1948, to complain that 'the Army was in a parlous condition, and was in a complete state of unreadiness and unpreparedness for war'.

It was not as if the armed services had nothing to do. Far from it: peace might have brought an end to the fighting but it had not produced any lasting stability. Within weeks of the end of the war in Europe the Allies found themselves confronting their erstwhile friends, the Soviet Union. British troops were involved in clashes with communist forces in Yugoslavia and Greece and the situation in Germany worsened, with the creation of West Germany, which was allied to the West, while communist East Germany became the front line of the Soviet and Warsaw Pact countries. In response to that threat a defensive alliance, the North Atlantic Treaty Organisation (NATO), came into being in 1949 with the British Army of the Rhine (BAOR) allotted to the Northern Army Group under the command of British generals. The confrontation was largely a war of words and a test of nerve, but it led to the creation of huge armoured and air forces on both sides of the border, while the threat of all-out nuclear war was never far away, and at times very real indeed. The confrontation lasted until the 1990s, when communism collapsed in Eastern Europe as a result of the acceptance there of the free-market economy and an unwillingness to tolerate totalitarian regimes.

At the same time, Britain had to protect its world empire. India was the biggest holding and the one which demanded most military support. It also underpinned the eastern empire, but in

1947 Attlee's government bowed to international pressure and to inescapable economic facts – the US disapproved of the imperial connection and the cost of maintaining India was becoming insupportable. Independence was granted and India and Pakistan came into being in August 1947, but the altered strategic balance did not stop Britain pursuing an aggressive policy in the Middle East to protect the vital sea route through the Suez Canal. At various times between 1946 and 1967, huge and vastly expensive garrisons were maintained in Egypt, Palestine, Libya, Cyprus and Aden. Not only were these costly to the exchequer but they were also unpopular and, according to Correlli Barnett in his history of the British Army, the British soldier had to bear the consequences as 'a dreary pattern repeated itself – murder, arson, ambush, cities divided by wire, road blocks, searches'. And it was not just in the Middle East that regiments saw operational service. Between 1947 and 1967 the army was engaged in operations in Aden, Kenya, the Gold Coast, British Honduras, Singapore, British Guiana, Hong Kong, Nassau, the Cameroons, Jamaica, Kuwait, Zanzibar, Borneo, Tanganyika (later Tanzania), Uganda and Mauritius.

Against that strategic background the KOSB began its service in the post-war world. The 1st battalion had ended its war in north Germany but it was soon on the move to a new deployment in Palestine, where Britain was attempting to keep the peace between rival Zionist and Arab groups prior to the formation of the state of Israel in 1948. Standing between the two sides were the 20,000 men of the Palestinian police force, backed by 80,000 soldiers of the British Army's 1st Division (1st KOSB's parent division), 6th Airborne Division and, from March 1947 when it was deployed in southern Palestine, 3rd Division. It was a deployment which Britain was hard-pressed to make and throughout the period the military units involved in internal security duties were severely overstretched. Inevitably, fighting an invisible enemy had a demoralising effect on the members of the British security forces. They were able to disrupt terrorist activities and in some cases forestall them, but all too often they found themselves reacting to situations about which they had no prior knowledge. As the *Borderers Chronicle* noted 'every man's hand seemed to be against us' and young Jocks learned to trust neither Jew nor Arab. This bred a siege mentality best expressed by General Sir John Hackett, who was more blunt, asking in Naomi Shepherd's history of the uprising why 'three thousand third-class Jewish lunatics could incarcerate and render impotent the flower of the British Army'. With the emergence of the state of Israel in May 1948, 1st KOSB and the other British regiments left the Holy Land, travelling overland by Gaza into Egypt, never to return.

For the 2nd battalion there was a different and much shorter period of soldiering. It stayed on in India and was one of the British regiments which oversaw the transfer of power from British Indian rule to the creation of the independent countries of India and Pakistan. During the deployment the battalion was based at Peshawar in the North-West Frontier Province, which became part of Pakistan. Most of its duties were in support of the civil power, mounting internal security operations and helping to keep the peace during the run-up to independence. For a regiment with long connections with the sub-continent and that particular area, it was fitting that the 2nd battalion should have ended its existence in Peshawar: as part of the post-war cutbacks to the infantry it was disbanded in 1947.

For the Territorial battalions there were also changes. The 4th and 5th battalions remained intact until 1961, when they amalgamated to form 4/5th KOSB, while the wartime 6th and 7th battalions were amalgamated, respectively, with the 4th and 5th battalions in 1947.

By then the regiment had settled down to its post-war existence, with one Regular and two (later one) Territorial battalions. The other big change was the introduction of post-war National Service. Wartime legislation for conscription was kept in place and under a succession of National Service Acts it became the law of the land for every male citizen to register at his local branch of the Ministry of Labour and National Service as soon as he became 18. Information about the relevant age-groups and precise instructions were placed in the national newspapers and broadcast on BBC radio, and schools and employers also played their part in passing on the relevant official information to their young charges. Short of deliberately refusing to register, there was no way the method of call-up could be ignored and those who did try to avoid conscription were always traced through their National Health records. Between the end of the war and the phasing-out of conscription in 1963, 2.3 million men served as National Servicemen, the majority in the army. In its final form the period of conscription was two years, following two earlier periods of 12 and 18 months and, like every other regiment, The King's Own Scottish Borderers benefited from the contribution made by men who were the first peacetime conscripts in British history. Basic training was undertaken at a number of centres, including the regimental depot at Berwick-on-Tweed.

KOREA: 1951–52

Within three years of returning to Britain in 1948, 1st KOSB was once again facing a hot war in a far-flung part of the globe. This was the conflict in South Korea, where British servicemen had been sent in September 1950 to support a US-led United Nations (UN) army following the invasion of the country by its northern neighbour. Korea had been annexed by Japan in 1910 and had remained a Japanese colony until 1945, when the country had been split into two halves along the 38th parallel, the north becoming a communist regime and the south a hastily organised democracy. Although the UN entertained hopes that the two Koreas might be reunited in the future, the opposed regimes were antagonistic to each other's existence and to the artificially created boundary which divided them. Any idea that they might find a means of living in harmony was shattered on 25 June 1950, when North Korea invaded its southern neighbour. Shocked by the abruptness and unexpected power of the attack, the US successfully persuaded the UN to oppose the invasion – its argument in favour of armed intervention was helped by the absence of the Soviet Union from the Security Council in protest at the UN's refusal to recognise Communist China. The US acted quickly: General Douglas MacArthur, Commanding General of the US Army in the Far East, was despatched from Japan to appraise the situation and by the end of July the Americans had four divisions in South Korea. Although the US task force had command of the air and the sea, it was powerless to halt the North Korean advance and by August the UN forces were desperately defending the Pusan perimeter, their last line of defence in the south-east of the country. The line held, but as the war progressed, and Chinese 'volunteers' came to the assistance of North Korea, the reasons for the UN involvement blurred and the fighting became a stalemate of entrenched positions and artillery barrage as the rival armies fought it out along the 38th parallel while the politicians argued about ways of resolving the conflict. In July 1951 a Commonwealth Division was formed and a year later 1st KOSB was ordered to join it, forming 28 Brigade with 1st Royal Australian Rifles and 1st King's Shropshire Light Infantry.

The British servicemen who saw action in Korea between 1950 and 1953 had to fight their way over a barren, remote and barely known peninsula which was almost as far away across the globe as it was possible for them to travel. Much of the fighting was reminiscent of the trench warfare which their grandfathers had encountered on the Western Front; the weather was usually awful, freezing cold in winter yet hot and wet in summer, and they faced an enemy who asked for, and offered, no quarter. The most common reaction of the men of the Commonwealth Division, which provided three British, Canadian and Australian infantry brigades to the UN forces, was bewilderment: bewilderment about the identity of their enemy and bewilderment that they should have been fighting the war in the first place.

Some 20,000 British servicemen fought in Korea and although they were smaller in number than the American contribution, their role was not without importance. A number of RAF pilots flew on attachment with Commonwealth or American units, three squadrons of Sunderland flying-boats flew reconnaissance sorties from their bases in Japan, ships of the Royal Navy patrolled the Korean coastline to prevent infiltration and a carrier task force provided air cover for the battle ashore. The main contribution, though, was provided by the Army: in all, 16 infantry battalions served in Korea, backed up by four armoured regiments and eight regiments of artillery with engineering, ordnance, transport and tactical support. It was a hard, bruising war with high casualties: 71 officers and 616 other ranks were killed in action, 187 officers and 2,311 other ranks were wounded and 52 officers and 1,050 other ranks were listed as 'missing', of whom 40 officers and 996 other ranks were prisoners of war and eventually repatriated. The final KOSB casualties were 31 killed, 90 wounded and 20 missing. By comparison, 33,000 US lives were lost and the total UN casualty list was 447,697 officers and men killed or wounded in action.

By the time 1st KOSB arrived, under the command of Lieutenant-Colonel J. F. M. Macdonald, a flurry of peace talks had suggested the possibility of a truce, but by autumn the political stalemate had again degenerated into armed hostility. In October

28 Brigade took up new positions beyond the Imjin, successfully dislodging the Chinese from the gaunt slopes of the hills known as Kowang-San and Maryang-San. The ridges formed an arrowhead piercing into the enemy lines, and it was there that the men of 1st King's Shropshire Light Infantry (KSLI) and the KOSB faced some of the fiercest fighting of the war, at the beginning of November. Following a series of probing attacks the Chinese turned the full weight of their assault on the KOSB positions on the Maryang-San ridge. It was there that Private William Speakman, a Black Watch Regular attached to the Borderers, won a Victoria Cross after leading one attack after the other on enemy groups on Hill 217. His conspicuous heroism and powers of leadership stiffened the resolve of B Company, and against all the odds the Chinese assaults were beaten off sufficiently to allow an orderly withdrawal from the ridge.

To the right of the ridge the Borderers' positions were held on a feature called the Knoll by two platoons from C and D Companies, which had come under the command of 2nd Lieutenant William Purves following the wounding of the senior subaltern. Formerly a bank clerk from Kelso in the Scottish Borders, Purves was a National Service officer with a reputation for being quiet and retiring, but his effective defence of the Knoll allowed his men to retire in good order in the early hours of the morning on 4 November. According to the *Borderers Chronicle*, every wounded man from the two platoons was withdrawn from the ridge, and despite his own wound Purves personally oversaw the dangerous operation until all his men had been shepherded safely off the hill:

> The position on the Knoll held by 2nd Lieutenant Purves's platoon of C Company and 2nd Lieutenant Henderson's of D Company was obscure owing to communications being destroyed. However, at about midnight it was established

> that this gallant party was still holding out though running short of ammunition, that and Lieutenant Henderson being wounded, 2nd Lieutenant Purves had assumed command of the two platoons. Surrounded on three sides, the plight of this party was serious. 2nd Lieutenant Purves was ordered to try and fight his way out towards D Company who were now holding out on 'Peak', a foothill of Point 317. This unpleasant operation was brilliantly carried out under the very nose of the enemy on Point 317. This exploit was rendered even more remarkable in that 2nd Lieutenant Purves succeeded in evacuating all the wounded and the complete equipment of his platoon under heavy mortar fire.

For his conspicuous gallantry and devotion to duty 2nd Lieutenant Purves was awarded the DSO, a unique and never-to-be-repeated achievement for a National Service officer. (Later he returned to civilian life and became a distinguished banker, crowning his career as chairman of the Hong Kong and Shanghai Banking Corporation.)

The bravery shown by men like Speakman and Purves – and by the entire battalion – allowed 1st KOSB to complete a successful evacuation from the ridge at a low cost in lives. Seven Borderers had been killed and 87 wounded, but after the battle it was estimated that the Chinese 'human wave' tactics had led to their sustaining around 1,000 casualties. At one point the men of D Company had run out of ammunition and been forced to hurl rocks and even beer bottles at the advancing enemy; on a more serious logistical note, the KOSB Mortar Battery fired a total of 4,500 rounds during the night-long engagement. Both the KSLI and the KOSB spent another eight months in Korea, fighting through a winter which saw the war along the 38th parallel degenerate into a struggle of static defence reminiscent of the Western Front battles during the First World War. From this period onwards the domination of No

Man's Land became the crucial test of the war as the stationary armies probed each other's defences. Hardly a night passed without infantry patrols infiltrating the Chinese lines, testing the enemy's strength, collecting intelligence information and attempting the difficult task of capturing Chinese prisoners for interrogation. (This latter duty reaped few rewards as most Chinese and North Koreans managed to commit suicide before – and in some cases, even after – being taken.) Most of the patrols were led by young subalterns, many of them National Servicemen.

MALAYA: 1955–58

The battalion left Korea in August 1952 and following stays in Scotland and Northern Ireland (Ballykinlar) it was sent first to Singapore and then to Malaya in August 1955. By then the British Army was in its seventh year of the period known as the 'emergency' which saw it involved in counter-insurgency operations against communist terrorist fighters (known as CTs) of the Malayan Races Liberation Army (MRLA), which was the military wing of the Chinese-controlled Malayan Communist Party (MCP). The emergency had begun in 1948, when heightened political tensions led to the MCP being declared illegal and some 10,000 MRLA fighters moved into the jungle to mount guerrilla operations against the civilian population and the security forces under their military commander, Chin Peng. Initially the idea was to drive the terrorists into the jungle away from urban populations but this changed in April 1950 with the appointment of Lieutenant-General Sir Harold Briggs as Director of Operations 'to plan, co-ordinate and to direct the anti-bandit operations of the police and fighting services'. To achieve those ends Briggs integrated the efforts of the police and the military and reorganised the intelligence services to provide information about terrorist movements and to infiltrate the communist cadre infrastructure. A 'food denial' policy was also instituted, but the main obstacle was the support given to the MRLA by the Chinese inhabitants of the jungle. The solution was the resettlement of 650,000 villagers in 550 New Villages – secure areas where they would enjoy a safe and profitable environment away from MRLA influence. Then it was the task of the infantry to move into the jungle, to secure bases and drive the CTs into the deeper and less hospitable depths.

By the time 1st KOSB arrived Briggs had been replaced by General Sir Gerald Templar, who refined the tactics and introduced a stronger political element. His credo was expressed in a statement he used over and over again: 'The shooting side of the business is only 25 per cent of the trouble, the other 75 per cent is getting the people of this country behind us.' The MCP had launched insurgency to gain independence for Malaya, but successive federal and state elections were also held from 1952 with the aim of achieving that ambition. This policy encouraged moderate politicians, but as the communists were excluded from the process they were gradually sidelined and independence was eventually granted in August 1957. Even so, military operations continued unabated, with a new emphasis on deep jungle patrolling and the introduction of psychological warfare with intensive propaganda to undermine MRLA morale. During the battalion's three-year deployment, mostly in north Johore, its rifle companies were heavily involved in jungle operations, a demanding existence which required stamina and endless patience on the part of the men involved. An article written for the *Borderers Chronicle* by 2nd Lieutenant A. D. S. MacMillan, a National Service officer, gives some idea of the conditions:

> The jungle night falls quickly, and trees and bushes assume menacing forms which move and threaten the longer one looks at them. You sit with your rifle across your knee and

> gradually the senses become taut as you strain into the darkness. The jungle floor is patterned with the luminosity of decaying leaves. You can lay a pattern of these to guide your hand to anything you may require during the dark hours. The brain becomes a sort of telephone exchange receiving and sorting out the myriad sounds into their respective origins.

It was not until 31 July 1960 that the 'emergency' in Malaya was declared to have ended.

HOME AND ABROAD: 1958–92

The battalion returned to Britain in September 1958 before undertaking its first operational tour in post-war Germany as part of the garrison in Berlin. While these deployments allowed 1st KOSB to recover from the excitements of two tough tours of operational duty in Korea and Malaya the period was not without its excitements. In 1960 and 1961 the battalion rugby team won the Army Cup, a notable achievement, as the battalion had not appeared in the tournament since 1938 and in both finals the opponents were 1st Duke of Wellington's Regiment, which prided itself on its rugby prowess. There was a rumour, not altogether unfounded, that the Dukes made sure that its regular intakes of National Servicemen contained excellent rugby players, many of them internationalists; not that the Borderers were shy of parading their own native talents – one of the battalion's star players was Brian Shillinglaw, a tough and talented scrum-half who won five Scottish international caps in 1961 playing for Gala and would have won more had he not changed codes to play rugby league. (Another great KOSB player was Frank Coutts of the Melrose club, who served in the 4th battalion during the Second World War and won three international caps in 1947. A stalwart of army rugby for many years, he later became President of the Scottish Rugby Union and retired from the army in 1973 in the rank of brigadier.)

In 1961 the regiment was affected by an unusual occurrence which showed both the strength of regimental traditions and a lack of understanding of that ethos by civilians. While any regiment is on home service it is a common practice for the Officers' Mess to hold cocktail parties to entertain local members of the community and friends of the regiment. At one such gathering in Redford Barracks in the summer of 1961 three guests removed the regiment's colours from the mess following a party thrown by the battalion. The incident was widely reported and the colours were subsequently returned by one of the men concerned, who also made a full written apology. So seriously did the regiment treat the incident that the commanding officer paraded the battalion so that he could address them and read the culprit's admission of guilt:

> We have served in many places where it was necessary to guard our colours night and day. We did not think this would be necessary in our own country. But, on Saturday night, three men took advantage of our hospitality and removed the colours from their rightful place. I am convinced that they did not appreciate the gravity of their act. They looked on it as an enterprising prank. They had not served in a famous infantry regiment and they did not appreciate this was an act of sacrilege.

The theft and subsequent return of the colours was front-page news (the commanding officer's words were fully reported in the *Scottish Daily Express*, at the time a newspaper which took a great deal of interest in military matters), and one of the culprits was obliged to resign his position on the Glasgow Stock Exchange.

A year later 1st KOSB was on the move again when it was ordered to deploy to Aden to take part in internal security duties following Britain's decision to establish its headquarters of Middle East Command in the colony, which was due to achieve independence in 1968. However, at the same time the position had been destabilised by an Egyptian-backed coup in neighbouring Yemen in September 1962, shortly after the battalion arrived. This led to an outbreak of violence in Aden and the declaration of a state of emergency as the terrorists escalated their attacks on government targets. Although 1st KOSB's tour of duty ended in 1964 it was called back to the area a few months later to join 'Radforce' (45 Commando, 1st Royal Scots, 1st East Anglian, 3rd Parachute Regiment, plus armour and air support) for operations against insurgent tribesmen in the Radfan, the mountainous border region adjacent to Yemen. The trouble had started with the creation of the British-sponsored Federation of South Arabia, a combination of the colony of Aden and the 20 various sheikhdoms and emirates in the area, to prepare them for independence. In January 1964 three battalions of the Aden Federal Army (FRA) had attempted to put down a tribal revolt in the Radfan which had been aided and abetted by the Arab National Liberation Front (NLF), itself sponsored by Egypt, but it had become quickly bogged down, hence the need for reinforcements. The campaign in the Radfan was a real test; not only was the topography harsh and the weather conditions unpredictable, but the men faced a hidden enemy who used a wide variety of tactics similar to those used by Pushtun tribesmen in earlier conflicts in Afghanistan and the North-West Frontier Province. Against that the British regiments enjoyed air superiority and could call on helicopters for supply and re-supply.

Following a short tour in Scotland which was used to organise a number of recruiting drives and a return to Shorncliffe as part of the United Kingdom Strategic Reserve, the battalion was moved once more to the Far East in the summer of 1965. Trouble had broken out three years earlier over the future of Britain's three remaining colonies in the area – Sarawak, Brunei and Sabah, all of which constituted British Borneo. The prime minister of Malaya Tunku Abdul Rahman wanted to include them in a new Malayan Federation but this was opposed by President Ahmed Sukarno of Indonesia, who wanted to incorporate the colonies into a greater Indonesia. The first trouble was in Brunei at the end of 1962, but although the rebellion against the sultan was crushed with British support, Sukarno opened a new offensive which became known as the 'Borneo confrontation'. There was a rapid escalation in the violence along the 970-mile land frontier and by the time 1st KOSB arrived the British contingent in the region had grown to 13 infantry battalions, one battalion of Special Air Service regiment, two regiments each of artillery and engineers, 40 strike aircraft, and 80 helicopters as well as local police and border security force units. Throughout the operation the tactics were similar to those which had been used in Malaya in the previous decade, but in this case greater use was made of helicopters to dominate the jungle. Patrolling these inhospitable areas dominated the battalion's tour of duty, which came to an end in the following year after Sukarno was removed from power on 11 August 1966 in a military coup.

A year later 1st KOSB moved to Osnabruck in Germany to face a fresh challenge as a mechanised battalion equipped with the new FV-432 tracked fighting vehicle, capable of carrying a section into battle. The conversion programme was a challenge, but the battalion rose to it and quickly settled down to the BAOR operational year. At the end of 1969 there was a further change, when it was deployed for the first emergency tour of Northern Ireland in support of the Royal Ulster Constabulary. This was in answer to the request made by the government of Northern Ireland in August 1969 for the provision of troops to assist the civil power in restoring order following outbreaks of sectarian violence in Belfast and Londonderry. Between then and 31 July 2007, when Operation Banner finally came to an end, Northern Ireland was almost a second home. Each tour brought its own challenges in helping to keep the peace and maintain a sense of proportion in one of the most difficult and long-lasting counter-insurgency wars fought by the British Army. The main opponents were the Provisional Irish Republican Army but trouble was also fomented by unionist terrorist groups and other troublemakers. As Robert Woollcombe made clear in his regimental history, extracts from 1st KOSB's log during its first tour in Northern Ireland from May to September 1970 give some idea of the complexity of the situation and the need for a balanced response from the security forces:

> Under no circumstances were the troops to open fire unless under persistent petrol bomb attack, and then only by marksmen at specific targets after many warnings . . . C Company snatch squad has gone in . . . a priest states he cannot talk reason to hooligans in the Flax Street area . . . We are being stoned, most of the street lights are out, but the Electrolux is working wonders . . . A very drunken woman is being restrained at the butcher's shop: request for an ambulance and a policewoman . . . An explosion at a house in Donegal Park Avenue, front door blown in . . . Civilian vehicle standing outside docks, two girls aged nineteen found in the boot . . . A nightmare indeed.

The emergency tours of duty in Northern Ireland and deployments with BAOR were very much part of the battalion's way of life in the last two decades of the twentieth century, but at the end of 1990 the world was plunged into a new crisis when President Saddam Hussein of Iraq ordered his army to invade Kuwait. This illegal action was followed by a lengthy game of diplomatic cat-and-mouse which culminated with the issue of the UN's Resolution 678, requiring Iraq to pull out of Kuwait by 15 January 1991. At the same time armed forces were deployed in Saudi Arabia and the Gulf region in preparation for offensive operations to oust Iraqi forces from Kuwait. It was the largest deployment by the British Army since the Second World War, and although the regiment was not employed in the offensive operations it did have a vital role to play. With 1st Coldsteam Guards and 1st Royal Highland Fusiliers it formed the Prisoner of War Guard Force (PWGF), which ran the main internment camp at Al Qaysumah to handle the Iraqi soldiers who fell into the hands of the UN coalition forces. During the operation the Iraqis were held in 'cages' – barbed wire enclosures with officers separated from other ranks in compliance with the Geneva Convention – but they also needed shelter. The response was to dig trenches and provide them with large sheets of brown plastic, but the gesture was not always appreciated as the Iraqi POWs thought that it was a prelude to their execution – a common occurrence during Iraq's earlier conflict with Iran (1980–88).

AMALGAMATION AND IRAQ: 1992–2006

Not long after leaving Iraq the regiment received the bombshell news that it was to be amalgamated with The Royal Scots, the oldest line-infantry regiment in the British Army. It was no secret that there were going to be cutbacks and amalgamations in the army, but the general impression was that regiments from the Gulf deployment would be spared. Like the KOSB, the Royals had served in Operation Granby and, again like the KOSB, it was one of only five line-infantry regiments never to have been amalgamated, the others being The Green Howards, The Cheshire

Regiment and The Royal Welch Fusiliers. (Later, all became part of larger regiments as, respectively, 2nd Yorkshire Regiment, 1st Mercian Regiment and 1st Royal Welsh Regiment.) The previous summer the government had produced its Defence White Paper, *Options for Change*, which proposed that the army should be reduced from 155,000 to 116,000 soldiers and that the infantry should lose 17 of its 55 battalions. The cuts came as a result of the end of the Cold War, following the disintegration of the Soviet Union and the reunification of the two Germanys. It was accepted that Scottish regiments would be affected but following both battalions' service in the Gulf the decision came as a shattering blow and immediate steps were taken by the regiment to fight it. Operation Borderer, a well-organised and high-profile campaign, was initiated and as a result the amalgamation was cancelled on 3 February 1993, together with the proposed amalgamation of the Cheshire and Staffordshire Regiments.

Alas, the reprieve was to be shortlived. Twelve years later, as a result of the Strategic Defence Review of July 2004, it was decided that the size of the infantry would be reduced still further and that amongst other changes The Royal Scots would amalgamate with The King's Own Scottish Borderers. Unsurprisingly, given the attachment which Scots feel for their regiments, the decision was not welcomed by everyone and determined attempts were made to save Scotland's regiments. Just as the previous proposal had caused anger because it came at a time when 1st KOSB had been serving in the first Gulf War, so were the latest reforms announced after a difficult and dangerous deployment in Iraq following the campaign to unseat President Saddam Hussein in 2003.

Although the actual fighting phase of the operation was successful, the country rapidly succumbed to a vicious insurgency war which needed the deployment of large numbers of troops. The British Army was responsible for security in the southern sector of

Basra, and between May and November 2003 1st KOSB was based in Maysan Province, where it had the task of attempting to restore stability while under constant attack from Iraqi dissidents. Once again it seemed that a historic and highly regarded regiment was bearing the brunt of unnecessary defence cuts and would lose its identity as a result.

This time the change was even more radical and far-reaching, as it involved a comprehensive restructuring of the infantry. Under these changes the size of the infantry was reduced from 40 to 36 battalions, and that meant the end of the remaining 19 single-battalion regiments. In their place nine large regiments consisting of several battalions were formed; in Scotland this new formation was called The Royal Regiment of Scotland, and the amalgamated Royal Scots and King's Own Scottish Borderers formed its 1st battalion. Formation day for the new regiment took place on Tuesday, 28 March 2006, the anniversary of the formation of The Royal Scots, the oldest antecedent regiment in the new large regiment. Four months later, on Tuesday, 1 August 2006, the anniversary of the Battle of Minden, The Royal Scots and The King's Own Scottish Borderers Battalions of the Royal Regiment of Scotland merged to form The Royal Scots Borderers, 1st battalion The Royal Regiment of Scotland. As 1 SCOTS the battalion continues to recruit from the parent regiments' traditional recruiting areas – Lothians, Borders, Dumfries and Galloway and Lanarkshire (the latter inherited from The Cameronians after that regiment's demise in 1968) – and many aspects of the KOSB uniform were retained (the drummers wear Leslie trews). Both Regimental Councils worked hard and in harmony to maintain the 'golden thread' which binds the new regiment to its historical antecedents, but that could not disguise the sorrow which many Borderers felt when 317 years of proud service to Crown and country finally came to an end on Minden Day, 2006.

Appendix

REGIMENTAL FAMILY TREE

1689: Leven's Regiment, The Edinburgh Regiment
1751: 25th (Edinburgh) Regiment of Foot
1782: 25th (Sussex) Regiment of Foot
1795: 2nd battalion raised and disbanded
1804: 2nd battalion re-raised
1805: 25th (King's Own Borderers) Regiment of Foot
1816: 2nd battalion disbanded
1859: 2nd battalion re-raised
1881: 1st and 2nd battalions York Regiment, King's Own Borderers
1887: 1st and 2nd battalions The King's Own Scottish Borderers
1947: 2nd battalion disbanded
2006: The Royal Scots Borderers, 1st battalion The Royal Regiment of Scotland

REGIMENTAL BADGE

The Cross of St Andrew upon a circlet inscribed 'King's Own Scottish Borderers'. Within the circlet Edinburgh Castle with three turrets, each with flag flying to the left. Above and below the circlet, within scrolls, the *In veritate religionis confido* (I trust in the faith of my belief) and *Nisi dominus frustra* (In vain without the Lord). Surrounding the circlet, a wreath of thistles. The Royal Crest with St Edward's Crown surrounds the whole. The crest is popularly referred to as the 'Dog and Bonnet'.

REGIMENTAL TARTANS

Until 1882 the regiment dressed as an infantry regiment of the line and did not wear tartan. Following the Cardwell/Childers reforms, in common with all Lowland regiments, it wore trews of Government or Black Watch tartan. In 1898 the wearing of the Leslie tartan was permitted and was issued to all ranks six years later. The regimental pipers wore kilts in Royal Stuart tartan.

REGIMENTAL PIPE MUSIC

Pipers were not officially recognised by the army until 1854, when all Highland regiments were allowed a Pipe-Major and five pipers. Before that most Highland regiments employed pipers as a regimental expense and these were distributed throughout the regiment disguised on the muster roll as 'drummers'. Lowland regiments followed suit. The pipes and drums were always fully-trained infantry soldiers and were in addition to the military band, which existed until 1994.

The regiment's pipe music is regularised as follows:

March in quick time: Blue Bonnets o'er the Border

March in slow time: The Borderers

The Charge: The Standard on the Braes o' Mar
A Company: The Bugle Horn
B Company: Bonnie Dundee
C Company: The Mucking o' Geordie's Byre
Support Company: Liberton Polka
Headquarter Company: Cock o' the North
Admin Company: Caber Feidh
Command Company: The Barren Rocks of Aden

BATTLE HONOURS

Two colours are carried by the regiment, the King's or Queen's which is the Union flag and the Regimental Colour (originally First and Second Colour), which is blue. In the centre, on a crimson background, is the regimental badge surrounded by the name of the regiment, and encircled by a wreath of thistles, roses and shamrocks, Also emblazoned on the Regimental Colour is a Sphinx superscribed with the word EGYPT, the Royal Crest in the first and fourth corner with the motto *Nisi dominus frustra* and in the second and third corners the White Horse with the motto *Nec aspera terrent (*Undaunted by adversity).

During the Napoleonic wars battle honours were added to the colours. In their final form, those gained during the First World War and the Second World War are carried on the Queen's Colour and the remainder are carried on the Regimental Colour. At the outset battle honours were given sparingly or even randomly. In 1882 the system of battle honours was revised by a War Office committee under the chairmanship of General Sir Archibald Alison. It laid down guidelines whereby only victories would be included and the majority of the regiment had to be present. Additional refinements were made in 1907 and 1909 and their recommendations form the basis of the regiment's pre-1914 battle honours.

Pre-1914 (25th)

Namur 1695	**Martinique 1809**	**Tirah**
Minden	**Afghanistan 1878–80**	**Paardeberg**
Egmont-op-Zee	**Chitral**	**South Africa 1900–02**

After the First World War there were further refinements to take cognisance of the size and complexity of the conflict. It was agreed that each regiment could carry ten major honours on their King's Colour but supporting operations would also receive battle honours which would not be displayed. The battle honours in bold type are carried on the Queen's Colour.

The First World War (12 battalions)

Mons	Le Transloy	**Hindenburg Line**
Le Cateau	Ancre Heights	Epehy
Retreat from Mons	**Arras 1917, 1918**	Canal du Nord
Marne 1914, 1918	Vimy 1917	Courtrai
Aisne 1914	Scarpe 1917, 1918	Selle
La Bassée 1914	Arleux	Sambre
Messines 1914	Pilckem	France and Flanders 1914–18
Ypres 1915, 1917, 1918	Langemarck 1917	
Nonne Bosschen	Menin Road	Italy 1917–18
Hill 60	Polygon Wood	Helles
Gravenstafel	Broodseinde	Landing at Helles
St Julien	Poelcapelle	Krithia
Frezenberg	Passchendaele	Suvla
Bellewaarde	Cambrai 1917, 1918	Scimitar Hill
Loos	St Quentin	**Gallipoli 1915–16**

Somme 1916, 1918
Albert 1916, 1918
Bazentin
Delville Wood
Pozières
Guillemont
Flers-Courcelette
Morval
Lys
Estaires
Hazebrouck
Kemmel
Soissonnais-Ourcq
Bapaume 1918
Drocourt-Queant
Rumani
Egypt 1916
Gaza
El Mughar
Nebi Samwil
Jaffa
Palestine 1917–18

In 1956 it was agreed to treat the Second World War in the same way. Those in bold type appear on the Queen's Colour.

The Second World War (16 battalions)

Dunkirk 1940
Cambes
Odon
Cheux
Defence of Rauray
Caen
Esquay
Troarn
Mont Pinçon
Estry
Aart
Nederrijn
Arnhem 1944
Best
Scheldt
Flushing
Venraij
Meijel
Venlo Pocket
Roer
Rhineland
Reichswald
Cleve
Goch
Rhine
Ibbenburen
Lingen
Dreierwalde
Uelzen
Bremen
Artlenberg
North-West Europe 1940, 1944–45
North Arakan
Buthidaung
Ngakyedauk Pass
Imphal
Kanglatongbi
Ukhrul
Meiktila
Irrawaddy
Kama
Burma 1943, 1945

Post 1945 (1st KOSB)

Kowang-San	**Korea 1951**	**Gulf 1991**
Maryang-San		

ALLIED AND AFFILIATED REGIMENTS

Canada

1st battalion The Royal New Brunswick Regiment (Carleton and York)
2nd battalion The Royal New Brunswick Regiment (North Shore)

Australia

25th/49th battalion The Royal Queensland Regiment

Malaysia

5th battalion The Royal Malay Regiment

South Africa

Witwatersrand Rifles

WINNERS OF THE VICTORIA CROSS

Lieutenant G. H. B. Coulson, 1st KOSB, Boer War 1901

The regiment's first Victoria Cross was won by Gustavus Hamilton Blenkinsopp Coulson at Lambrechfontein on 18 May 1901 whilst serving in the mounted infantry role. Seeing a fellow trooper in difficulties Coulson rode to his rescue but both men were thrown when the horse was wounded. Coulson ordered the man to save himself by riding to safety and he himself was almost saved by another corporal but they were both gunned down by heavy Boer fire. Coulson was born in Wimbledon in 1879.

Piper Daniel Laidlaw, 7th KOSB, First World War, 1915

Daniel Laidlaw, the 'Piper of Loos', was one of a select band of Scottish pipers who were awarded the Victoria Cross for continuing to play their instruments while under heavy enemy fire. Shortly before the attack on Hill 70 Laidlaw's battalion came under a massive artillery bombardment which caused casualties and dented morale. Ordered to play his pipes, Laidlaw climbed onto the trench parapet in full view of the enemy and played the regimental charge, 'The Standard on the Braes o' Mar'. During the fighting he was wounded but he survived the battle and returned to his native Berwickshire after the war. He died at Shoresdean near Berwick-on-Tweed in June 1950 and is buried at nearby Norham.

CSM John Kendrick Skinner, 1st KOSB, First World War, 1917

A native of Pollokshields in Glasgow, Sergeant Skinner was awarded the Victoria Cross for his courage and determination under heavy enemy fire at Wijdendrift in Belgium. Finding that his company was pinned down by machine-gun fire he collected six men and attacked the three enemy blockhouses, bombing and capturing the first on his own and leading his men to successfully attack the other two. He was killed in March 1918 trying to bring in a wounded man, and had the distinction of having six VC holders as his pall bearers.

CQMS William Henry Grimbaldeston, 1st KOSB, First World War, 1917

Another Victoria Cross was won at Wijdendrift in similar circumstances to the award made to Sergeant Skinner. In this instance Sergeant Grimbaldeston also gathered a party of men to attack a German blockhouse. In his case he managed to work his way to its entrance, where he threatened the machine-gun team with being bombed. As a result 36 Germans surrendered and six machine guns and a trench mortar fell into the battalion's hands. A native of Blackburn, Lancashire, Grimbaldeston died there in 1959.

Sergeant Louis McGuffie, 5th KOSB, First World War, 1918

The only KOSB Territorial to be awarded the Victoria Cross, Sergeant McGuffie came from Wigton in Galloway. During a battalion attack on Piccadilly Farm near Wytschaete in Belgium he attacked and captured a number of enemy dug-outs and took a large number of prisoners. During the same action he prevented the Germans from leading away some British soldiers who had been taken prisoner. Less than a week after the action he was killed by a German shell.

Private William Speakman, 1st KOSB, Korean War, 1951

One of four Victoria Crosses awarded during the Korean War. Speakman was a native of Altrincham in Cheshire and served originally with 1st Black Watch. Having volunteered for service in Korea, he was in action with 1st KOSB during the action on the Maryang-San Ridge, where his conspicuous heroism and powers of leadership stiffened the resolve of those around him. The fighting lasted six hours and when the men ran out of ammunition Speakman encouraged them to continue the fight with rocks and stones. Against all the odds the Chinese assaults were beaten off sufficiently to allow for an orderly withdrawal from the hill. Badly wounded during the incident, Speakman later retired from the army and lived for a time in South Africa.

Bibliography

Unless otherwise stated, extracts from soldiers' letters and diaries are in the possession of the regiment or are housed in the Imperial War Museum or the National Army Museum, London. Battalion and brigade War Diaries and other official papers quoted from are housed in the National Archives, Kew.

BOOKS ABOUT THE KING'S OWN SCOTTISH BORDERERS

Baggaley, Captain J. R. P., *The 6th (Border) Battalion, The King's Own Scottish Borderers 1939–1945*, Martin's Printing Works, Berwick-upon-Tweed, 1946

Blockwell, Albert, ed. Maggie Clifton, *Diary of a Red Devil: By Glider to Arnhem with the 7th King's Own Scottish Borderers*, Helion, Solihull, 2005

Coutts, Major F. H., ed., *War History of the 4th (Border) Battalion, The King's Own Scottish Borderers*, privately published, 1945; *One Blue Bonnet: A Scottish Soldier Looks Back*, B&W Publishing,

Edinburgh, 1991; *The Golden Thread: Mair Tales from the Brig*, B&W Publishing, Edinburgh, 2006

Dolbey, Captain R.V., *A Regimental Surgeon in War and Prison*, John Murray, London, 1917

Gillon, Captain Stair, *The KOSB in the Great War*, Thomas Nelson, London, 1930

Goss, Captain J., and others, *A Border Battalion: The History of the 7th/8th (Service) Battalion King's Own Scottish Borderers*, T. N. Foulis, Edinburgh, 1920

Gunning, Captain Hugh, *Borderers in Battle: The War Story of The King's Own Scottish Borderers 1939–1945*, Martin's Printing Works, Berwick-upon-Tweed, 1948

Higgins, Captain R. T., *The Records of The King's Own Borderers or Old Edinburgh Regiment*, Chapman & Hall, London, 1873

Macdonald, Major-General J. F., *The Borderers in Korea*, Martin's Printing Works, Berwick-upon-Tweed, n.d.

Richardson, Gavin, *For King and Country and the Scottish Borders: The Story of the 1/4th (Border) Battalion The King's Own Scottish Borderers on the Gallipoli Peninsula*, privately published, 1987; *After Gallipoli: The Story of the 1/4th (Border) Battalion The King's Own Scottish Borderers 1916–1918 Egypt, Palestine and the Western Front*, privately published, 1992

Scott Elliot, Captain G. F., *War History of the 5th Battalion King's Own Scottish Borderers*, Robert Dinwiddie, Dumfries, 1928

Tullett, Captain E.V., *From Flushing to Bremen: The 5th Battalion, The King's Own Scottish Borderers*, privately published, 1945

Weir, Rev R.W., *A History of The Scottish Borderers Militia*, Courier & Herald, Dumfries, 1873; *The History of the 3rd Battalion, King's Own Scottish Borderers*, Courier & Herald, Dumfries, 1918

White, Peter, *With the Jocks: A Soldier's Struggle for Europe 1944–45*, Sutton Publishing, Stroud, 2001

Woollcombe, Robert, *Lion Rampant: 15th Scottish Division from*

Normandy to the Elbe, Chatto & Windus, London, 1955; *All the Blue Bonnets: The History of The King's Own Scottish Borderers*, Arms and Armour Press, London, 1980

OTHER BOOKS CONSULTED

Ascoli, David, *A Companion to the British Army 1660–1983*, Harrap, London, 1983

Barnett, Correlli, *Britain and her Army 1509–1970*, Allen Lane, London, 1970; *The Lost Victory: British Dreams, British Realities 1945–1950*, Macmillan, London, 1995

Baynes, John, with Laffin, John, *Soldiers of Scotland*, Brassey's, London, 1988

BBC website http://www.bbc.co.uk/ww2peopleswar/

Blake, George, *Mountain and Flood: The History of the 52nd (Lowland) Division 1939–1945*, Jackson & Co, Glasgow, 1950

Bloem, Walter, *The Advance from Mons 1914, The Experiences of a German Infantry Officer*, Peter Davies, London, 1930

Brereton, J. M., *The British Army: A Social History of the British Army from 1661 to the Present Day*, The Bodley Head, London, 1986

Chandler, David, and Beckett, Ian, eds, *The Oxford Illustrated History of the British Army*, Oxford University Press, Oxford, 1994

Churchill, Winston S., *A History of the English Speaking Peoples*, Vol III, Cassell, London, 1957

Connor, Kevin, and Gootzen, Har, *Battle for the Roer Triangle: Operation Blackcock – January 1945*, privately published, 2006

Ewing, John, *History of the 9th (Scottish) Division 1914–1919*, John Murray, London, 1921

Fortescue, Sir John, *A History of the British Army*, 13 vols, Macmillan, London, 1899–1930

Henderson, Diana M., *The Scottish Regiments*, Collins, Glasgow, 1996

Holmes, Richard, ed., *The Oxford Companion to Military History*, Oxford University Press, Oxford, 2001

Jackson, Bill and Bramall, Dwin, *The Chiefs: The Story of the United Kingdom Chiefs of Staff*, Brassey's, London, 1992

James, Colonel Lionel, *The Indian Frontier War: Being an Account of the Mohamud & Tirah Expeditions 1897*, Heinemann, London, 1898

Keegan, John, *Six Armies in Normandy*, Jonathan Cape, London, 1982

MacKay of Scourie, Lieutenant-General Hugh, ed. J. M. Hogg, P. F. Tytler and A. Urquhart, *Memoirs of the War carried on in Scotland and Ireland, 1689–1691*, Edinburgh, Bannatyne Club, 1833

Middlebrook, Martin, *Arnhem 1944: The Airborne Battle*, Viking, London, 1994

Mileham, P.J.R., *Scottish Regiments*, Spellmount, Tunbridge Wells, 1988

Neillands, Robin, *A Fighting Retreat: The British Empire 1947–1997*, Hodder & Stoughton, London, 1996

Pearce, Nigel, *The Shield and the Sabre: The Desert Rats in the Gulf 1990–91*, HMSO, London, 1992

Royle, Trevor, *The Best Years of Their Lives: The National Service Experience 1945–1963*, Michael Joseph, London, 1986

Shepherd, Naomi, *Ploughing Sand: British Rule in Palestine 1917–1948*, John Murray, London, 1999

Stewart, J. and Buchan, John, *The 15th (Scottish) Division 1914–1919*, William Blackwood, Edinburgh, 1926

Strawson, John, *Gentlemen in Khaki: The British Army 1890–1990*, Hutchinson, London, 1989; *Beggars in Red: The British Army 1789–1889*, Hutchinson, London, 1991

Trench, Charles Chenevix, *The Frontier Scouts*, Jonathan Cape, London, 1985

Wood, Stephen, *The Scottish Soldier*, Archive Publications, Manchester, 1987
Younghusband, Sir George, *The Relief of Chitral*, Macmillan, London, 1895

Index

INDEX

INDEX

INDEX

INDEX